BUT ENOUGH ABOUT ME

BUT ENOUGH ABOUT ME

A JERSEY GIRL'S UNLIKELY ADVENTURES
AMONG THE ABSURDLY FAMOUS

Jancee Dunn

HarperCollins*Publishers*

The names and identifying characteristics of some individuals were changed to protect their privacy.

HarperCollins books may be purchased for educational, business, or sales promotional use. For information, please write: Special Markets Department, HarperCollins Publishers, 10 East 53rd Street, New York, NY 10022.

FIRST EDITION

Designed by Kris Tobiassen

Printed on acid-free paper

Library of Congress Cataloging-in-Publication Data is available upon request.

ISBN-10: 0-06-084364-0

ISBN-13: 978-0-06-084364-9

06 07 08 09 10 NMSG/RRD 10 9 8 7 6 5 4 3 2 1

I'm nobody! Who are you?
Are you nobody too?

—EMILY DICKINSON

BUT ENOUGH ABOUT ME

How to Jolly Up a Surly, Hungover Band During an Interview

Approach with caution. Often the band members have only recently arisen, even if it is well into the afternoon. Do not be cheerful. Avoid openers that sound parental, such as "Well! Looks like someone had a late night!" Have some breath mints handy, in case one of them has recently thrown up or has neglected to brush his teeth. Oral hygiene is not very "rock," so be prepared.

As soon as the group is settled and their handlers have scurried to dispense energy drinks and aspirin, immediately name-check their tuba 'n' bass concept album that was released only in Germany, so that they know that you Get Them. Inevitably, however, your exhaustively researched questions will produce grunted, monosyllabic answers, for the band members will not want to seem like some eager teen pop group. Their goal is to make music, they will tell you pointedly, not to bone chicks or make videos or have their drinks paid for or stay in plush hotel rooms. Thus it is their duty to convey that these interviews are a nuisance, and they would be just as happy rehearsing in a garage somewhere. At this time you must roll out the heavy artillery. *Pay attention only to the drummer.* Laugh uproariously at his jokes. Stare with dumbfounded awe as he offers up his philosophies. Shake your head and say things like "I never thought about it before, but you are absolutely right—drumming *is* a metaphor for life!" Listen, rapt, as he explains to you the genius of John Bonham's skinsmanship. As the puzzled but excited drummer blossoms under your admiring gaze, his other band mates, particularly the heretofore-mute sunglasses-wearing lead singer, will at first be confused, then annoyed. Finally, their competitive spirit will take over and they will enthusiastically jockey for attention, offering amusing anecdotes about groupies and telling off-color jokes.

Do not use any quotes from the drummer.

1.

I am fifteen. I am going to my first concert unaccompanied by my parents. This is thrilling for a number of reasons. One, because I was invited by Cindy Patzau, the most glamorous girl in my sophomore class, still glinting with stardust after a recent performance during a school assembly in which she did a dramatic interpretive dance to Cyndi Lauper's "Time After Time." She wore a clingy black bodysuit in front of the whole school. She was my hero.

"You want me to go with you?" I squeaked when she called. I sat with the popular kids in our high school cafeteria, but I certainly wasn't A-list. When I got my braces off that year, no one noticed for a week, whereas when Liz Kincaid had hers removed, there was much squealing and jubilation in the halls. During senior year I was voted Class Clown when I desperately wanted Best Legs (won by a girl with the movie-star moniker of Jill Shores). As the clown, I was the peripheral Don Rickles figure to the bronzed, care-free Dean Martins and Frank Sinatras, bristling with sour flop sweat, one bad joke away from being banished from the Sands. At the time of Cindy's call, I was on unsteady social ground due to a recent gaffe at a party. I was leaning against a wall, waiting in the bathroom line, when a senior named Mark, a hip soccer player who wore Adidas Sambas and liked the Clash, materialized behind me. He smirked. "Holding up the wall?" he asked.

Tell me, what is the sharp, snappy rejoinder to "Holding up the wall?" I gawped at him as everyone in the line nudged each other, waiting for my trademark lightning comeback. Holding up the wall. Holding up the wall. Seasons passed. The leaves on the trees outside withered, dropped, bloomed, and withered again. Holding. Wall. Mark abruptly turned away from me and started chatting up another girl. Good-bye, Rat Pack, hello, dinner theater in Jupiter, Florida.

"This show," I said to Cindy. My words came out in a high-pitched, phlegmy squawk: *Zhis gghow.* I hurriedly cleared my throat. "Is it just you and me?" Surely there would be others.

"Yes," Cindy said calmly. "I know you have good taste in music, so the ticket won't be wasted." While I was processing this, I heard the click of a phone being picked up in my parents' bedroom. It was my younger sister Dinah. I could tell by her breathing. If I didn't play this phone call right, it could be my Waterloo, and I was frantic that Dinah shouldn't hear any bumbling. I needed to scare her. I inched toward the hallway in order to get a view of the bedrooms upstairs. Because there were three girls in our family, the phone cord in our kitchen had been stretched until it was ten yards long in our efforts to have a little privacy. Recently, my youngest sister, Heather, had managed to reach the hall closet, and conducted her preteen business with the door shut and key words muffled by the coats. I stretched the cord, gently but firmly, and crept over to where I could just glimpse Dinah in my parents' room. I waved furiously and her head jerked up. *Goddamn you,* I mouthed, affecting a tough squint. She froze like a snowshoe hare—out of fear, or stubbornness, I couldn't tell—but she didn't hang up.

While I fought rising hysteria, Cindy detonated this: The concert was to take place at a college. We would have to cross the New Jersey state line to Haverford College in Pennsylvania. With her older sister! And we'd spend the night! In a dorm room!

"Cool," I said, elaborately casual. "I'm in." I could hear Dinah's sharp intake of breath. She knew as well as I that it would take a typhoon of tears to persuade my strict father to let me go. *Hear me out, old man,* I thought grimly (he was thirty-nine at the time). *I am going. Oh yes. I am going.*

A week later, after frenzied negotiations with my parents that rivaled the SALT talks in length and intensity, I was allowed to accompany Cindy to Haverford. The night before I left, after a bout of gastrointestinal distress at the thought of hanging around a VIP like Cindy for a sustained length of time (this would become a lifelong pattern), I retired to my room to pack.

Soon enough, there was a timid knock on the door. Dinah and Heather stood silently, knowing that they must be invited in. "Hey, can we watch you pack?" asked Dinah. At fifteen, I still held powerful sway over my younger sisters, and I carefully polished my mystique. Usually when they were allowed to enter the sanctum, it was so that I could extort their cash. My "garage sales" were a frequent scheme. "Garage sale in my room, five o'clock," I would announce briskly as they raced to their rooms to scrounge for money or begged the folks for a forward on their allowance. Meanwhile, I rummaged through my drawers for tchotchkes to unload: a frayed collection of Wacky Pacs, a half-empty bottle of Enjoli, a trio of black rubber Madonna bracelets. As they waited by the door, twitching with eagerness, I would build momentum by popping my head out every once in a while with updates. "Five more minutes," I'd bark. "Lotta good stuff in here, lotta good stuff. I really shouldn't be selling some of this." Finally I would fling open the door and they would push over each other, running.

During one of these bazaars, my mother watched from the doorway, arms folded, lips pursed. "You should be ashamed of yourself," she said.

"Why?" I asked coolly, shutting the door on her. "For bringing color and excitement into my sisters' lives?"

I also gave various lessons. Ballet instruction cost fifty cents, seventy-five cents for the deluxe. For that particular con, I recited instructions into a tape recorder ("Point your toe forward, and back; repeat"). When my two customers arrived, I pressed "play" and walked out, only feeling a twinge later when Heather said, "I wish you hadn't left. We were so disappointed because we wanted to be with you."

Another proven revenue stream was music-appreciation seminars. "Now, do you two remember who this is?" I'd say, carefully putting *Crimes of Passion* on the record player as they sat cross-legged on the floor.

Heather frowned. "Blondie?" she ventured.

"Pat Benatar," I'd say crisply, pacing back and forth. "This is called 'Treat Me Right.' Pat's from Long Island. She used to be a waitress. She is going out with her guitarist, Neil Giraldo. Got it? Dinah, are you taking notes?"

This was one of the few times I was not interested in their cash. Still, I drew out the moment by continuing to pack as they waited on the other side of the door. "We've got cookies," Heather called. "I just made them. Sugar cookies." Taken from an old recipe in an ancient *Better Homes and Gardens* cookbook, sugar cookies were a family staple, blindingly white thanks to Crisco, white flour, and cups and cups of sugar. Eagerly, I opened the door. They bounded over to my bed and we all flopped onto it, shoveling warm cookies into our mouths. Then, after our sugar high spiked, I got down to business, imperious once more. "I need to pick out an outfit for the show," I said, rising from the bed and opening my closet doors. What they would never guess is that as my back was turned to them, I was thinking, *I wish that it could always, always be just like this, with you two giggling and jostling each other on the bed next to my elderly, snoozing cat. Must I leave? Must you leave? Can't we stay?*

"You look good in everything," said Heather loyally as I held up a pair of elasticized aqua paisley In Wear pants. Heather was five years younger than I, so she was an easy sell. Two years separated Dinah and me, so her compliments were less effusive.

"What about pajamas?" I fretted. I couldn't wear my pink flannel Lanz nightgown to a college.

"Wear your good sweatpants," Dinah counseled. Only in New Jersey could you have "formal" sweatpants. "And if you promise to take care of it, I'll give you my Hard Rock London T-shirt."

"Slippers?" I said.

She shook her head. "No. Socks, but not dark ones. Peds would be best."

"What if some college guy tries to hit on you?" asked Heather gleefully, bouncing on the bed.

That was not going to happen, particularly since I had recently gotten a perm that was extreme even by mideighties New Jersey standards, rendering my hair as dense and impenetrable as a boxwood hedge. But of course, I couldn't tell them that.

"I'll do what I usually do," I said. There was no "usual." "I'll say I have a boyfriend."

After I shooed my sisters out of my room, I sat at my desk and wrote out a list of potential topics to bring up with Cindy in case there was dead space.

1. Do you think Mr. Boone looks better without his mustache?
2. What's your take on scrunch socks?
3. Do you watch *All My Children*? If so:
    a. Do you think Jenny and Greg make a good couple? If not, why not?
    b. Do you want to come over to my house during the big episode when Jenny and Greg get married? You can't? Oh, you have field hockey practice? No, right, it was a bad idea anyway.

The next day, exhausted after a nerve-jangling ride to Pennsylvania ("Do you know," I said to Cindy in one of my many conversational misfires after I had run through the list, "that you're a dead ringer for that actress Hedy Lamarr?"), we finally lined up in front of the concert hall. Slowly I came to life. My first show! The college kids around us all looked impossibly poised. How do they know where to buy clove cigarettes? Did they do those asymmetrical hairdos themselves, or did they go to a salon? After an eternity, we made it through the door. I took a deep breath, tasting the gloriously stale, loamy, cigarette-tinged air.

"Let's try to get up close," Cindy said. I was afraid of crowds but I had to impress her, so I charged recklessly through the audience so we could secure a position near the front. Guitar techs in scraggly ponytails, shorts, and tube socks darted around the stage, adjusting equipment. The crowd began to cheer. Then: Out went the lights. My pulse surged crazily. *I could do this every night,* I thought, ping-ponging with excitement. *Every night of my life.*

And then, a spotlight came on, the band bounded onstage, and an announcer hollered those life-changing words: "*Folks! Please give it up! For! THE HOOTERS!*"

Mastering the Crucial Opening Patter

When you first meet a celebrity for an interview, remember one thing above all else: Do not, under any circumstances, work yourself into the opening chitchat. Want a dead gaze and a rigid smile? Start off the proceedings by saying that you're a "big fan." Stee-rike one! Never, ever lead with the word *I*. Mention at your peril that their debut album prevented you from slitting your wrists during that "bad patch." You may think you are offering up a heartfelt gift, but most artists have been told this very tidbit hundreds, sometimes thousands of times. If their album was the sound track to any major event in your life—wedding, prom, and the like—by all means keep it to yourself. Unless the event was losing your virginity, which might work as an icebreaker for certain metal bands.

Keep in mind that you have only a minute or two to engage your subject, so those first few moments are key for celebrity-nobody relations. Famous people are like taffy: They are only pliant for a short period of time before they harden, and you're left with canned answers as their eyes flick around the room, seeking a rescuing publicist. It's a critical period, so you must avoid the classic pitfalls. Never lead off with any sort of flattery ("You look incredible! Were you just on vacation?"), which is too obvious and used too liberally by everyone in a star's orbit, so do not join the chorus. Flattery, for them, is the aural equivalent of soothing background noise, like the low murmur of a TV.

Never list the ways in which your subject looks different in person, which is conversational quicksand. Blurting out that they appear younger or thinner up close is a well-meaning but disastrous opener that not only starts the cogs of insecurity whirring in their minds ("Does this mean I don't look good on film?") but is something that they constantly hear from fans when they are stopped in the street. You must avoid reminding them of a fan, or else their expressions will calcify into a bland, ever-so-slightly exasperated game face right before your eyes. In nine out of ten instances, if your subject is an actor, he or she will also be shorter in person than they appear onscreen: This, also, you must keep to yourself. Even if you think you are giving their lack of height a positive spin, you aren't. "You always seem larger than life in photos, but it's nice to see that in person you're just like us" might seem like a compliment, but what a star hears is *You're stumpy, and you will lose jobs to taller people.*

If your subject is a musician, do not offer your "take" on their lyrics. And never compare their music to other bands' because it is an absolute no-win (if the competing band is superior, they will feel anxious; if it's perceived as substandard, they're insulted). If the band is a one-hit wonder, carefully avoid mentioning the hit, because you, of course, are interested in their more obscure, cruelly overlooked work. If you happen to find yourself interviewing E.U., best not mention "Da'Butt"; if Wall of Voodoo is your subject, skim over "Mexican Radio."

If your subject is older than you, avoid pointing out that their work is a great favorite of your parents, or that one of their albums was the very first one you ever bought when you were in junior high. If it's thunderingly obvious that your subject's best work is well in the past, do not mention any classics from an earlier album, or they will wince. Keep it in the present, even if an artist's latest output is on a tiny Internet record label run by their cousin.

Now that you know what to avoid in that critical first minute, how do you swiftly capture the attention of someone who is inured to both flattery and sincerity? You must surprise them with a Fun Fact about themselves. If you blow in with a newsy little item about *them,* there is instant festivity. Your celebrity can yell the news to their assistants (who are always nearby, usually talking at a low volume in the next room). They will come running in, and the party begins.

Scan your celebrity research as if it were the Dead Sea Scrolls, trolling for any fresh news or obscure nugget that might have escaped the eye of their handlers. Have they been mentioned in a media studies course at a university? A veneer of intellectual respectability is always a plus. Were they recently praised by a fellow celebrity? That's a can't-miss. Even better, create your own logrolling. If you have an interview on Monday and a different one on Wednesday, ask Monday's famous person if they're a fan of Wednesday's. Invariably the answer will be an inoffensive yes, and then you can pass along the warm tidings of admiration.

Before I interviewed the cast of *Friends,* the cabdriver who picked me up at the airport was listening to Howard Stern, who, as it happens, was dissecting the latest *Friends* episode. I took notes, knowing that the cast was taping the show that morning and wouldn't be listening. Then, when I met them for dinner, I rolled in, sat down, and regaled them with Stern's impressions of the show. Howard Stern is a sure thing, because people will always have an opinion about him, and we started the proceedings on a lively note, with everyone talking animatedly at once.

Anything is better than a stilted "How are you?" as you unpack your tape recorder while your bored subject waits, silently and expectantly. "How are you?" is wasted time, a meaningless exchange, because your celebrity will dutifully answer "fine" or "great" and then you are back to square one. When I was to meet Britney Spears in her dressing room during a rehearsal for *Saturday Night Live* in New York, I brought along a press kit that had been sent to me of some young blond singer who was billed as "the new Britney Spears." What you want to have happen is for her to call over her handlers, and for everyone to excitedly gather around and make the appropriate remarks, which is exactly what occurred. What fun is a mini-event unless you have your handlers materialize? Britney yelled to her assistant, her mama's best friend, Felicia, who was just returning to the dressing room with a snack for Britney, a large, bright pink strawberry milkshake from McDonald's. A stylist ran over and made mean comments about Britney's hapless mini-me. Good times!

Fun Facts can arrive serendipitously. Before heading to Nashville for a chat with Dolly Parton, I was at the gym reading *Harper's Bazaar* on the treadmill when I came across an interview with Donatella Versace. She was asked whom she would love to dress, and she mentioned Dolly. Jackpot! *Harper's Bazaar* had just come out that day, and I took a gamble that the couture bible might not be Dolly's required reading.

Although an old-school pro like Ms. Parton didn't really need buttering up, I reasoned that it couldn't hurt. When we made our introductions, I imparted my Donatella Fun Fact. "What?" she hollered. "Well, how about that! Her clothes are probably tacky enough for me, right?" She called in her assistant to tell her the news, and the hand-clapping mood was set.

If all else fails, surf eBay and hunt for kooky merchandise that relates to your celebrity. At the very least, you can breeze in and open with, "Did you know that the bidding for one of your cigarette butts is up to twelve dollars and fifty cents?" Plop down chummily into a chair and continue. "And one of your sweaty tank tops is up to forty, and no reserve." A normal person might recoil at your stalkerlike tendencies, but most famous people will hear this and light up like Times Square.

The process of engaging your celebrity is not unlike being a photographer at the Sears portrait studio. You just need a different version of a squeaky toy so that their eyes follow you and they smile occasionally.

2.

My path toward interviewing the famous was a meandering one. When I was growing up, I loved being at home with my family in my small New Jersey town and certainly had no intention of ever leaving. With the myopia of youth, I assumed that every family across the nation shared our little foibles. Our cuisine, for instance: Surely, everyone on earth relished the fiber-free beige food that my family loved to bolt down—crescent rolls from a can, boil-in-a-bag noodles. Is it puffy? Is it off-white? Pull up a chair! Crab dip, French toast, twice-baked potatoes! Garlic bread, buttered Uncle Ben's rice, fettuccine Alfredo, turkey on white! Don't forget the mayo, or the Ex-Lax, for that matter, because you won't be going to the bathroom for the duration of the weekend! A sound that will always make me mistily nostalgic is the *fsst! fsst!* of I Can't Believe It's Not Butter being squirted on a biscuit. Ah, oui, à la recherche du temps perdu.

One of our favorite savory beige treats—reserved especially for holiday times—was called "breakfast strata," in which you pour eggs over stale white bread, add a pound of cheese and another pound of greasy crumbled sausage. Bake in the oven, and gobble as you talk around the table with your mouth full. When I would beg my southern mother to give me an after-school snack, she would do the following: take a slice of bread, slather some margarine on it, and dump a heapin' helpin' of sugar on top. Fold it over, and voilà! A sugar sandwich. It was heaven, the sugar crystals crunching between my teeth.

It was only when I served one of these delectations to some girls in my sixth-grade class that I realized that my family was a little different. The girls, Paige and Jennifer, were my social betters, and I had coaxed them over to my house after school, serial-killer style, with the promise of playing with a new kitten and eating unlimited snacks. "Um, what is this?" asked Paige, wrinkling her nose as I breezily handed her a sugar sandwich.

"Oh, does it need more sugar?" I asked, lifting up a corner of the bread.

Paige and Jennifer looked at each other. "You go ahead," said Jennifer freezingly. I saw Paige ease her Docksider shoe over to Jennifer's and nudge it, ever so slightly. My chance at trading up cafeteria tables at lunch was over.

My mom was markedly different from the benign, fluttery suburban mothers of my friends. She hailed from the tiny town of Citronelle, Alabama, and had a slight southern accent. Of course, when I imitate her to my sisters, she becomes a dyspeptic Foghorn Leghorn: " 'Yoh *fahthuh* and I were out on the *veranduh*' " (they don't have a veranda, they have a deck) " 'enjoyin' a jeulip an' some peppuh cheese strahws as we gazed out over the *proh*-puh-ty.' " The key is to accent every fifth word, and to throw in every hackneyed reference to the South that you can think of: magnolias, hoop skirts, Sherman's march.

No kid in my town had a mother who was an erstwhile beauty-pageant winner. Mom was the very first Oil Queen of Citronelle. It was exotic for my friends to see the yellowing photo of Judith Ann Corners from 1960, holding a spray of roses, a crown shaped like a little oil derrick perched gaily on top of her head. When she was crowned at the Citronelle High School auditorium, Mom was presented with a seventy-five-dollar prize and a visit with Citronelle's mayor, who had the *Wizard of Oz*–sounding name of B. L. Onderdonk.

As befits a former beauty queen, Mom was always chic and well turned out. I don't think I've ever seen her without lipstick, and always, always a bright pedicure, usually cherry red. Even when she was lounging around the house, she was wearing something trim and colorful, as opposed to my frighteningly random getups of ratty pajamas or breezy, carefree combos: undies 'n' T-shirts, robe 'n' slipper socks, sports bra 'n' sweats. When we were kids, she carefully dressed us up for airplane trips in order to "get better service."

The prototypical steel magnolia, my mom was eccentric enough to keep things interesting. I knew, instinctively, that other kids weren't threatened with colorful southern expressions like *I'm going to slap you upside that wall.* Once, after a summer visit to Citronelle, Mom noticed that her brother Leslie Ray and his wife, Jackie, would have our cousins Thad and Tray (both names pronounced the southern way, with two syllables) answer every question with "Yes, sir" or "Yes'm." This didn't translate very well in New Jersey, but my mom was obsessed, and so we were forced to give it a go. After dinner, we were instructed to chirp, "I enjoyed it, excuse me, please." Thanks, Thad and Tray! Heather was in preschool at the time and didn't really grasp the whole "etiquette" thing. "'Joyed it," she would lisp insanely, bringing her sausage breakfast strata to the table. I ask you: How was I supposed to have any friends in middle school when we were told to recite, "Dunn residence, this is Jancee speaking," when the phone rang? She and my father were strict, using fear as the time-honored disciplinary tool: of being grounded, of extra chores, or simply of them being displeased.

Particularly my mother. Let's say that I was getting a little sassy at the dinner table. As I mouthed off, my mother would be calmly chewing her food, staring down at her plate as she munched her chicken-fried steak thickly coated with Club cracker crumbs. Once she decided that she had had enough, she would fasten her eyes on me with a bored expression, which carried an ever-so-slight undercurrent of menace. Slowly she would raise her fork and point it at me, tines down. For a few terrifying seconds, she would hold it there, while my feisty words died in my throat and I gawped at the fork the way the mongoose sees the cobra. Then it would drop, and she would calmly continue eating.

We three kids got our drive squarely from her. The woman had grit. After a stint at Auburn University, she got restless and decided that the only way she could escape her small town and see the world was to become a stewardess. In those days, one of the job requirements was to have perfect vision, and she was practically blind. Undaunted, she made an appointment for an interview at United Airlines. When it came time to read the eye chart, she asked the instructor in a drawl as sweet as Karo syrup to fetch her a glass of *watuh.* She ran over, memorized the chart, and promptly landed the job.

From my father, we girls got our excessive sentimentality and our pathological love of habit. Here is a man who ate Raisin Bran—always Post, never Kellogg's—for decades ("Keeps you regular as a clock!"), who put timers on all the lights in our house so that they would switch on at six p.m. and off at eleven, and who, every Sunday through at least three presidential administrations, played the album *Gord's Gold,* from bearded Canadian troubadour Gordon "The Wreck of the Edmund Fitzgerald" Lightfoot. Here was a man so mushy that his record collection also contained a number of anthologies that he ordered from late-night television commercials with names like *Slow Dancing* (Paul Anka's "Puppy Love") and *Secret Love* (Air Supply's "I'm All Out of Love").

My father worked for JCPenney as a teenager all the way until his retirement, as did his father before him, and his father before him. "Over one hundred and seventy-five years of service," my dad liked to boast about our family (that, and the more terrifying "Not a single relative of ours has ever gotten divorced").

I have always thought of James Cash Penney as a kind of distant, benevolent uncle. My father's name is J. C. Dunn, as was his father's before him, and I was supposed to carry on the family legacy as J. C. Dunn the Third. When I surprised my folks by being female, my parents hastily cooked up the name "Jancee" (remember, it was the sixties). I have a cousin named Penny, too, who . . . worked for JCPenney. How I wish this were all a coincidence.

In the den, my father proudly displayed a photo of Penney's lunchtime visit to our grandfather's house. On a nearby shelf—near the book *Power of Integrity: J. C. Penney*—sat a small bronze bust of J. C. Penney's head, and really, wasn't he much more relevant to our lives than some dimly remembered ex-president or Roman ruler? Arrayed on every available surface were group photos of JCPenney district managers, grinning cheerfully, most with a glasses-and-mustache combo, all with good, solid names: Vern Leister, Dutch Koenig, Dave Gable.

My father conducted both his business and his family life by the Golden Rule, a set of business principles established by Penney after he started his first store in 1908: Do unto others as you would have them do unto you. Treat people with courtesy and respect. Be honest. Dad had fervently hoped that one of us girls would work for Penney's and continue the family heritage, but it was not to be, aside from one summer when Heather did a brief, grudging evening-shift stint at the Penney's women's department at the Rockaway Mall, selling elasticized pants to older ladies before racing to catch a keg party. We always chafed at Penney's. As kids, we happily wore Garanimals, their children's line of easy-care pants and tops in seventies color schemes like striped avocado and orange, but as teens, we turned up our noses at Penney's offerings. We grew up in the preppy, status-obsessed eighties, and the items that we hungered for—Lacoste shirts, Tretorn sneakers, Bermuda bags—were out of our folks' price range and, more important, not offered at JCPenney.

"There are good versions of all those things at Penney's," my father would insist. He just didn't understand that while the Penney's renditions were skilled copies, they were still a tad off in crucial ways. Instead of Levi's, for instance, Penney's offered Plain Pockets.

"They don't have the red tag," Heather pointed out.

"Exactly," said my father. "Other than that, you can't tell the difference."

"*I* can tell the difference!" Heather raged, thumping up the stairs to her room and banging the door shut. "Okay? *My whole school can tell the difference!*"

"Bunk," said my dad.

Instead of Lacoste alligator shirts, Penney's offered the Fox. "It's half the price, and that's before my employee discount," my father said. "And it looks like an alligator from far away."

"I'm just going to save up my babysitting money and get an alligator shirt," said Dinah, handing him back the bag.

He also struck out when he brought home Penney's translation of the Polo shirt. It was called Hunt Club, but instead of Ralph Lauren's horse and rider, Hunt Club just had the horse. "There's no guy on top of the horse," I pointed out.

"Oh, for Christ's sake," my father said irritably.

Our youthful snottiness reached a pinnacle when Dinah and I were snooping in my parents' closet before Christmas and unearthed a Penney's bag of presents. "What are these?" Dinah wondered, pulling out some pink sweatpants and a sweatshirt. She read the label aloud. "Sweatworks," she said. I usually led Dinah into trouble. She was my compadre, always willing, when we were kids and playing pretend, to dress up as the monster. Or the prince. On snowy days, we would take a sled outside, attach it to a rope, and play Horse and Carriage, and Dinah never put up any fight about being the horse. When my father built Heather a dollhouse, Dinah spent the whole day decorating it as a surprise, down to the most intricate detail: carpeting, pictures on the wall, fake food on the table. She was the most softhearted girl I knew, and I manipulated her shamelessly.

"What the hell is that?" I asked, examining the Sweatworks. I was the "edgy" one in the family, so I tended to swear a lot. "They're kind of like Generra sweatshirts. Only they're not."

"They're so not," agreed Dinah.

I held them up to myself. "I guess I could cut off the sleeves." This practice was fashionable at the time.

"Or we can wear them inside out. I saw Tory Fair do that at school."

I rolled my eyes. "Sweatworks!" I said, holding them up to myself and pretending to walk.

Dinah joined in, cackling. "Oh, are you admiring my sweatshirt? It's from *Sweatworks*!" We pranced around the room.

"You mean you haven't heard of Sweatworks?" I cried, tossing a purple set into the air. I looked at Dinah to see if I was getting a laugh, but she was motionless, gazing, stricken, at the doorway.

My father was standing there. "I guess you found our Christmas presents," he said. He gathered up the Sweatworks and silently put them back in the bags. He suddenly looked tired, and sad. Dinah and I could barely meet each other's eyes. We were absolutely radiating shame, and when Christmas came, we put on those pink and purple Sweatworks right from the box and kept them on all day. "Dad, I seriously need sweats, this is great," I said with hearty cheer.

"Do these come in other colors, Dad?" asked Dinah. "I might need to get more."

The thing was, while I wouldn't wear Hunt Club shirts, I still loved JCPenney. I loved paging through the giant catalog, with its bland-looking models frozen in unnatural poses as they cavorted on a fake beach, or smiled contentedly as they pretended to sleep in a throw pillow–strewn bed. It soothed me to see the cheery, retro clothing categories ("Spring Separates, in Missy Sizes, Too!" "Easy Care Slacks with Comfort Stretch!" "Step into Slippers: Flirty, Floral, Feminine!"). As I thumbed through the home-merchandise pages, I was comforted by the Vidal Sassoon Hard Bonnet Hair Dryer, the Standard Toilet Lid Cover in Dusty Rose or Bronze Gold, the Cozy Recliner in Fashion Colors.

Best of all was visiting my father at the actual store in Wayne, New Jersey. My dad was the manager, and his employees loved their tall, handsome, affable boss—if they had a wedding, he was there; if a relative died, he showed up for the funeral—so they were especially kind to us girls. If we bought some Ultima II lipstick, a winking counter girl would toss samples of mascara and nail polish in the bag. As we rampaged through the Campus Shop or Home Expressions armed with his employee discount card, my father would make his eternal rounds through the store, wearing his natty suit with the "Jay

Dunn" nametag, joking with clerks but keeping a sharp eye out for anything that was amiss—a Separate that had slipped off a hanger, a gum wrapper that had fallen onto the rug. Well after his retirement, when we made our frequent stops to check out other Penney stores ("I heard this one does a big business in portable audio"), my father would constantly bend over to pick up any trash that he saw on the floor.

When he was on duty at the store, he would talk to every customer, to our everlasting horror. My dad had a handy repertoire of four or five all-purpose comments that he used on shoppers. A husband glumly trailing his wife? "I see she dragged you out here!" Or "I'll bet you'd rather be home watching the ball game!" A lady leaving the store laden with bags? "I see you bought us out!" "If the customer has kids, I'll say something about them, because people always like to talk about their kids," he said. "A little bit of humor is good." He waved to a store clerk. "With Penney's," he finished proudly, as though he were making a presentation, "it's about the first and last impression—someone in the store saying hello and good-bye."

Who doesn't love a good old-fashioned department store? Back in the day, Penney's boasted even more departments, because it strove to be your complete one-stop-shopping destination. The building in Wayne housed an automotive department, a garden shop, a candy pavilion, and a pet store. ("The snake got loose again," said my father one night, wearily pouring a Scotch. "And it ate some of the hamsters.")

The highlight of our spring was Penney's annual Sweet Sale. At the door, shoppers would receive a chocolate bar—the good, thick kind with almonds in them that high school kids sell—and tucked inside would be a coupon for either 25, 30, or 35 percent off of an item that day. "Another employee just slipped me a candy bar," Heather would gleefully report. After a few hours of shopping, we'd stop for cheeseburgers and cheesecake squares in the Penney's coffee shop, which, with its squeaky vinyl booths and soothing brown, tan, and mauve color palette, was one of our favorite places on earth to eat.

Penney's was the tamest of all of our family jaunts. Most of them were much wackier. One prized ritual was paying a carefree visit to the family cemetery plot, which my sensible father had purchased for all of us many years back. "We've got some property," he announced when the deal came through, as we gathered around him for details. "It's a double plot," he told my mother excitedly. "You could put four caskets in there, or we have enough for, oh, I'd say eight or ten cremations."

The plot came in handy when Heather and I were going through our Goth Lite phase (black clothing, major Maybelline eyeliner, silver jewelry). She and I would make late-night drives to the graveyard—blasting Siouxsie and the Banshees—so that we could drape ourselves moodily around our final resting place.

"You know what our friends are probably doing right now?" Heather said scornfully as she settled her back against a gravestone. "Watching TV."

I shook my head. "They just don't get it," I said. I looked up at the moon and sighed in a mournful way. Call me morbid, call me pale. "And you know what? They will never understand me in a million years. This is the real me. I'm not afraid of darkness, and I'm certainly not afraid of death." Actually I was, but I had to put up a good front because Heather was younger.

Heather absently pulled up some grave grass. "I wish that everyone in high school could see me right now. It's like they can't look beyond my Benetton outfits."

When the rest of the family got wind of our visits, they decided they wanted in on the action. Soon, we had a ritual. Someone would bring up our cemetery plot, and Heather would cry, "Let's make a night of it! Festively, we would pile into the car. First, we'd go to Fabio's Mexican Cantina in New Providence for gloppy, cheese-filmed Mexican food. ("Smile, and smile some more," the menu urged. "Laugh, and laugh some more.") Then we'd merrily drive off to our little patch of land.

"On a clear day," said my father, stepping out of the car, "you can see New York City from here. You can't beat the location."

"I dare you to lie down and cross your arms!" Dinah said to Heather. Soon, we were lying together on the plot, our arms folded like Nosferatu.

"Jay, where did you put the camera?" my mother cried. "Wait, I know, it's in the car! Dinah, stay there, with your arms folded just like that, and no smiling!" Each family member leaped up to take a photo of the others, while urging them to stay motionless. As my father was fiddling around with the flash, Heather whispered to me, "You know what? I don't think this is creepy, because when one of us does die, at least I'll have this memory and it won't be a sad thing." This was far too sunny a thought to be Goth Lite, but I liked it, anyway.

New York was only an hour-long drive away from our small town in Central Jersey, but my family visited the city exactly once a year, which was all the culture I required. We would have some moo shoo pork and egg rolls in Chinatown, then take the obligatory spin through Little Italy. For us kids, what was even better than eating a cannoli at Ferrara's was the ultimate in excitement: going to look at the bums on the Lower East Side, which we did faithfully once a year. It was kind of like visiting relatives.

"Let's go see the bums! Time for the bums!" we would chorus. Then my father would pilot our light blue Buick LeSabre downtown to the Bowery, which in the eighties was an open-air asylum, not the hipster carnival it is now. There we would park, and silently, reflectively watch the homeless men swill their Mad Dog 20/20. This was supposed to be a character-building, there-but-for-the-grace-of-God-go-I exercise, but for us kids, it was better than a matinee of *Cats.*

"I want you girls to think about how lucky you have it," my father would say as two muttering, shabby men broke out in a drunken fistfight.

"Jesus, Jay, lock the doors," my mother said, frowning anxiously.

Both of the men weakly put up their dukes, landing a few tepid body blows but basically just making a wobbly circle around each other. Nearby, a guy with a World's Greatest Grandma T-shirt was settling into a refrigerator box next to a jar half-filled with urine. As he gathered some newspaper around him, he squinted wearily up at us, a suburban family of five, staring wide-eyed at him from the car.

"It's easy to take things for granted," continued my father as Dinah nudged me. *Prostitute,* she mouthed. I craned my neck.

"—Simple things like a roof over your head. Electricity. A decent education—"

I raised my eyebrows at my sisters and jerked my head to the left. Two rats were in a fight to the death over what looked from the car to be part of a pastrami sandwich. Ugh, look at their long, pink tails. The bigger one started to gain ground. Whoops, now the little one is on top. Scrappy, that one.

"—valuable lesson for you about gratitude," he concluded.

We nodded meekly. Then my dad threw the Buick in reverse, and we headed to the Holland Tunnel, and back to New Jersey. The day was always exciting, but we all felt a little relieved when we pulled onto the Turnpike.

The Setting: Why the Chilcotin Plateau of Central British Columbia Is a Capital Idea

If it is at all possible, avoid setting up your interview in New York or Los Angeles, because your famous person will invariably be overscheduled and distracted. In a different era, celebrities would occasionally grant a "personality profile" when they didn't have a movie or a TV show or an album to plug, but these days, they are told by their handlers that to give an interview without something to sell just looks needy and sad. So when it is time to push a product, the public relations team usually loads up their star with multiple interviews per day over the course of a week or two.

The deadliest occur in an airless conference room of the movie company at their Los Angeles or New York headquarters. If this happens, all you can hope is that the chat takes place in the morning, when the star is freshest, rather than at the end of the day, when the star is deranged from answering the same five questions about gaining weight for a role or "what it was like" to work with various costars. It can be especially awkward when a movie was made years ago and was shelved or delayed (this happens a lot). The star will have trouble unearthing the obligatory anecdotes from the set because he or she will have made three films in the meantime, and will end up repeating the same two stories for each visitor, and you are left with nothing.

With musicians, it's always wise to pick a tour stop where there are no media for miles. Often publicists will try to combine a photo shoot and an interview to get it over with, but the subject is always being pulled in different directions, so you must grab five frantic minutes as they get changed, or have their makeup done, or you shout questions over the noise of the hair dryer. If a star is uncomfortable or too distracted to come up with a clever answer to your question, she'll get her entourage to participate and turn the question on a member of her glam squad ("I don't know. Chrissie, did I ever want to quit the business?") so that the answer is unusable.

Thus, it is always a good idea during the negotiations process to fish around and see if your subject is going to leave town for any reason. If he or she is a musician and they are on tour, pick a place like Kansas City for your interview. If he or she is a movie star and on location in a nonglamorous spot, try to wrangle your way in. Why? Because the star might actually be glad to see you. You will be a fresh new face, and if they are neurotic about being away from the public eye in a remote outpost, you will bolster their ego by conveying that the world still cares about what they are doing. Yes, a gang of screaming fans gathered by their hotel entrance can buoy them up, but a journalist who dramatically flies into a distant location is just that little bit more legitimate. The welcome will be warmer, the focus sharper, and if your celebrity is bored by seeing the same faces on tour or on set, you may get some extra time, rather than just the promised hour and a half.

The only roadblock becomes actually getting to the location. With Brad Pitt, in the Rockies for his mountain-climbing movie *Seven Years in Tibet,* I flew from New York to Vancouver, then on to the mountains in a twelve-seat Beechcraft terrifyingly named Wilderness Air, whose pilot spent the entire shaky barf bag of a flight with her head swiveled around, chatting animatedly with the passenger behind her. Her copilot, meanwhile, had his head buried in a book. So who, exactly, was flying the plane? The Lord? A computer? I knew that DC-10s had them, but Wilderness Air? I looked around for dibs on the meatiest-looking passenger to eat if the plane crashed.

Against the odds, we landed in a remote field and I wandered over to a diner, awaiting a van that was to pick me up for a three-hour journey up the muddy mountains. I supposed I should eat. I was constantly afraid of long stretches without food. As the waitress slapped a burger on my table, she gave me a hard look and said, "Been a lot of moose attacks around here lately. Mothers protecting their babies. They just come barreling out of the woods at ya." Then she shuffled away. Moose attacks? The only wildlife I had to worry about in my neck of the woods were squirrels, and rats. Which I hated. Yet I missed them.

After a long, vertiginous van ride with one stop to let a herd of caribou cross, I arrived at the far-flung mountain camp, one of the few places on earth where Brad Pitt could walk around unmolested. He was dressed for the six inches of mud that surrounded the camp, in boots, sweatpants, and a black suede coat. He was so friendly and positive, so free of attitude, that my palms were barely moist when I shook his hand. I was relieved that I didn't have a crush on him; this would make the proceedings go a little more smoothly. I was drawn to scrawny, tubercular indoorsmen, while Pitt was more of a hot ski instructor/beach bum type. His favorite expressions were "Yeah, right?" when he agreed with you on something, followed by "Yeah, man" (when he agreed with you on something but perhaps less stridently), followed by "Excellent." He referred to his costar, the British actor David Thewliss, as Thew-lie.

After I gamely followed him on a long hike, he invited me into his trailer. Yes! I quickly looked around, scribbling everything down on my pad: *Scientific American* magazine, a black Prada tote, a carton of Camels, a book on Frank Lloyd Wright, and a huge box of strawberry Twizzlers. Aha, a CD collection. The Dave Matthews Band, Shawn Colvin, and Soundgarden. As I got situated, Pitt decided to blast a few tunes from the Soundgarden album, and that is when it started to rain. In my palms. For as "Burden in My Hand" cranked up, Pitt began to rock out.

Chris Cornell's voice rang majestically over the mountains, exhorting us to follow him across the desert, as thirsty as you are.

Pitt thrashed around.

This has happened to me a couple of times, and I never know what to do. When someone is playing air guitar, do I play guitar as well? Whip my head around? Do I fill in on air drums? Pick up something and examine it? It's the same situation as being in the studio with a band and they play you some new tracks from their upcoming release. They all stare at you expectantly. Do you bob your head? Do you close your eyes, presumably lost in the music? You can't sing along, of course, because you never heard the words before. Maybe you give an OK sign and a big smile?

As Pitt leaped around the trailer and continued to rock, I stood there helplessly. Do I yell, "This album is so great"? I fast-forwarded ahead: He can't hear me, so he turns the stereo down, I idiotically repeat, "This album is so great," he is annoyed because I have cut his moment of abandon short. I opt to go to the trailer door and take in the spectacular view of the peaks. I assume an awed, but slightly mellow, expression to match his low-key manner. After a few more songs, he joined me at the doorway to remark that he never got tired of the view. I, meanwhile, exhaled in relief. And because this wasn't in Los Angeles, I was able to spend the entire day with him—hiking, eating lunch in the commissary, even watching *Sling Blade* in his trailer with him, just the two of us. It wasn't necessarily that I was spectacular company, it's just that the deep mud around the set had suspended the shoot, and he didn't really have anything else to do.

Sadly, most interviews do not have the option of taking place in the wilderness, but with a little ingenuity, even the most common of settings can work. Hotel rooms, for instance, are a music-world cliché, and not all that interesting unless they're filled with used crack vials, but if your rendezvous takes place in a hotel, all is not necessarily lost.

The San Antonio hotel room of Rolling Stones guitarist Ron Wood offered a perfect peek into his everyday life. He and the band were in the midst of a U.S. tour and I joined him during a night off. He opened the door to the coziest scene imaginable. Ron's blond wife, Jo, was bustling around doing the laundry and throwing out the occasional wisecrack. Candles were burning, scarves were draped over the lamps for a softer effect, and the TV was turned to an old interview with Katharine Hepburn ("Tough old broad!" Wood cackled appreciatively). My affable host, meanwhile, had set up a keg of Guinness beer in the bathtub of their suite and would periodically amble in there and fill 'er up. He urged me to join him, and he was so casual about it that after a while I unself-consciously headed to the bathroom to avail myself. As I filled my glass, I surveyed the surroundings. Draped over the shower curtain was a row of dripping black socks. Ron rinsed them out in the sink at night like an old British pensioner! This is a man who could have paid someone to lick them clean.

When I returned to the living room, Ron showed me a large sketchpad full of artwork that he had done, flourished here and there with an occasional guest doodle by Mick. We spent a delightful evening paging through the book and drinking Guinness while he reminisced about the Stones' early days, told stories about the band's visits to the homes of Jerry Lee Lewis and Fats Domino (whose wife wouldn't allow him to stay in the house, he had to sleep in a shed in the garden), and talked proudly about his kids. He was one of the most unaffected people I've ever met. At the end of the night, he led me over to the balcony of his room and threw the doors open to a Texas sky that seemed impossibly huge. A few nights earlier, he told me, he and Jo had come out there and spent a long while watching a spectacular electrical storm. "It was beautiful," he said quietly, gazing out at the stars.

3.

I gave up trying to escape my family long ago. We were like a boisterous pack of lemurs, twitchy and clannish, leaping frantically out of our den, turning en masse to the left, then to the right, chattering shrilly at intruders. When I graduated from the University of Delaware, I turned around and moved right back home, ostensibly to regroup. Perhaps "graduated" is sort of a broad term, because I was seventeen credits short of a degree. I was just desperate to escape Delaware, with its sprawling campus and its thick-necked frat boys, so I promised my father that I would attend summer school and make up the classes. My dad left Michigan State early to embark on his career at Penney's, so it broke his heart that I would continue the family legacy of being diploma-free. Of his three daughters, only Dinah has managed to graduate college, and no matter what I have achieved since I left Delaware, he periodically brings up those seventeen credits.

"It's never too late to get an education," he will say.

"It most certainly is," I reply. "One of my classes is a Biology with Lab. If I go to class at this point, I'll look like Rodney Dangerfield in *Back to School.*"

Then he inhales, and slowly blows out a breath through pursed lips, and I know that he's about to paint a Dark Picture. "Say you apply for a job, and they do a background check," he says. "They come to find out that you lied. You can kiss that job good-bye. What if you become famous—say you receive some award, something like that—and somebody decides to look into your past?" He shakes his head. "Kiss that award good-bye. Tell you right now."

"Who is this investigative team?" I ask him. "And why don't they have anything better to do? And what award am I getting, exactly?" He will not be distracted.

My father is the most genial midwestern guy imaginable, but for him, disaster lurks around every corner—financial ruin, squandered health, pyramid schemes, airbags failing to deploy—so he tends to use fear as a parenting tool to try to goad his daughters into being more prepared. This inevitably involves Kissing Things Good-bye. "Looks like mold," he'll say, standing up and brushing off his knees after inspecting the wood underneath Heather's porch. "Better get that sealed up, or you can kiss this porch good-bye."

Most often, it is your actual life that you can kiss good-bye. "Huh," he'll say when you tell him that no, you don't have a carbon monoxide detector. "Guess you didn't hear about that family in the news, down in Trenton. Went to sleep, never woke up. Some sort of problem with the oven. Carbon monoxide detector is twenty bucks. Your call." Then comes the Sorrowful Head Shake.

If he comes to visit, he will bring along trinkets that are designed to induce complete paranoia: cans of Mace that attach to your keychain, a plastic hood that filters out smoke in case you need to crawl out of your apartment during a fire ("Maybe you're not aware that most accidents occur at home"), a special doorjamb for use in a hotel room to prevent break-ins during the night, when you're most vulnerable. "You've got a constantly changing population in a hotel," he'll say grimly, handing over the unwieldy, hard-to-pack device. "You think a security guard can keep track of everybody?"

At the very least, he'll show up with a sheaf of papers from the latest *Consumer Reports,* the bible of preparedness, which he keeps in a file cabinet, indexed back to the Reagan administration. "Think it can't happen to you?" he'll say, pointing with his middle finger to a *Consumer Reports* article on long-term disability insurance. "Most companies are either cutting long-term disability or they never had it in the first place. A car hits you, you get paralyzed. How soon do you think your money is going to run out?" He stares at me, unblinking, his mouth a hard line. "Then what?"

I never know whether these are rhetorical questions, or if I'm supposed to answer him. Instead, his doomsday predictions fan my eternal flame of paranoia, which is always at a low burn. I imagine myself in an airless room, covered in a body cast, my arms stiffly protruding from my thin, dirty blanket. The only thing I can move is my eyes, which are fastened on my home health care aide—a recent parolee, because with my dwindling finances I can only afford someone in a work-release program. He is telling me he is quitting because I haven't paid him in weeks. "Don't make me mad," he hollers, throwing down my bedpan. "My girlfriend used to make me mad." I look on helplessly as flies gather on the pan to have lunch. Then they move to my face for an aperitif. "Please," I whisper weakly, through cracked lips. "Get them off my face." They crawl nimbly around my eyes.

"Bitch wouldn't shut her mouth!" he screams, upending my bedside tray.

The flies are laying eggs in my eyes.

My father studies my face, which is hypnotized with fear. Mission accomplished!

So it was with trepidation that he allowed me to enter the work force, where I would spend whole days away from his watchful eye. Armed with my partially completed degree in English, I attacked the want ads of the *Newark Star-Ledger* with enthusiasm. Hello, world! I spread out the paper on my folks' kitchen table and studied the ads, which glimmered with promise. I knew that I didn't want to be a teacher, the standard job assumption for an English major. Rather, my dream since I was knee-high was to be a writer for a newspaper or a magazine. As a kid, I devoured any magazine I could get my hands on. My mother's *Family Circle,* my father's *Time*—anything.

To me, the height of fun wasn't playing outside but pouring myself a big glass of cherry Kool-Aid and settling down with a nice, fresh issue of *Cracked* or *Mad.*

I had my first subscription at nine, a monthly for kids with the very seventies name of *Bananas,* available through the Scholastic Book Club at school. How I lived for the arrival of my beloved *Bananas*! All of my favorite stars graced its covers: Farrah Fawcett, Chewbacca, the guy who played Juan Epstein from *Welcome Back, Kotter* ("Bob Hegyes Is a Real Person!" read the cover line). I spent hours poring over articles like "Catching Up with *Logan's Run.*" There was an advice column, groovily titled "Good Vibrations," and a magic trick of the month offered by Magic Wanda, a kind of female Doug Henning who wore yellow overalls and signed her column "Love, peace and magic."

My very favorite feature was called "Bummers," a kvetch fest in which kids sent in complaints that began "Don't you hate it when . . ." If your grievance was sufficiently grating, it would run in the magazine, accompanied, excitingly, by an illustration. "Don't you hate it when your dog gets to stay up late and watch TV when you have to go to bed?" was a typical winner. I was brimming with complaints, even as a young child, so I feverishly composed pages of gripes to the "Bummers" desk. I was sure I had some hits. Don't you hate it when your mom makes you take a bath when you had one last week? Don't you hate it when your parents make you eat natural peanut butter instead of Jif like all of the other kids in the United States? Don't you hate it when you have to go to Colonial Williamsburg for summer vacation instead of Hersheypark, which has rides and a tour of a chocolate factory?

I never won, but it didn't dim my enthusiasm for *Bananas*—or, for that matter, Scholastic's other publication, *Dynamite.* I dreamed of working at *Mad* magazine, certain that it was a laff a minute. All I wanted for most of my life was to join the magical world of magazines. The problem, of course, was that I didn't have the faintest idea of how to go about it. During the summer of my senior year in college, I got my toehold with an internship at *New Jersey*

Monthly magazine, where I fact-checked articles on "The State's Best Subs." If I had more *cojones,* I would have then set my sights on a job in New York, the print media capital of the United States, but in my suburban Jersey bubble, New York seemed as far away as Canada.

Plus, I loved my home state. Ah, Jersey, God's country! New Jersey in the eighties was my lotusland. We lived in a preppy, upper-middle-class town with immaculate sidewalks, but at the end of the day, it was still situated in New Jersey (unofficial slogan: "Parts of it are nice"). If you're from Jersey, you can wear all the preppy clothes you want, but no one will mistake you for a Bostonian. You may have carefully built up a sophisticated veneer, but eventually your Garden State origins are going to surface like a herpes sore. Maybe your nails are just a millimeter too long, or a "Yeah, right?" slips out when you agree with someone. It could be the moment when your hostility rises after hearing the Giants maligned, or—my recent roots-affirming situation—when someone cut me off as I was driving on the Garden State Parkway listening to Bad Company on WDHA, "New Jersey's own rock station," while munching on a chicken parm sandwich. I sped up to tail them and screamed "Fuck you!" with my mouth full of half-chewed food.

Our town's ethnic mixture was Irish and Italian, and the town was occasionally rumored to be a Mob enclave. There were always whispers about the Badaraccos, the family that lived up our street. Frankie Badaracco was in my sister Heather's class. During recess one memorable day at her elementary school, Frankie was horsing around with a dirt bike. Somehow he managed to jam his finger in the spinning spokes of the wheel, where it sliced cleanly off and flew into the grassy expanse of the schoolyard. All of the kids, baffled and terrified by the sudden vulnerability of the loudmouthed Frankie, stood motionless, as if in a game of Freeze-Tag.

One kid managed to break free of his hypnosis and pedaled his bike frantically to the Badaraccos' house down the street. A few minutes later, a car roared up and Mr. Badaracco whipped out of the front seat, slamming the door. He strode toward us, holding up a crisp one-hundred-dollar bill between his thick fingers. "A hundred dollars for whoever finds my son's finger," he shouted hoarsely. The kids, relieved at having something to do, threw themselves into this gruesome version of an Easter egg hunt.

Years later, Heather was in a Hoboken bar, being chatted up by some guy.

"I'm from your hometown," he said. "I'm a few grades below you."

She squinted at him. "No you aren't," she said. "It's a small town. I would recognize you."

"I can prove it," he said. "I found Frankie Badaracco's finger."

What wasn't there to like about New Jersey? I never understood the jokes. I loved piling into my friend Janet's Chevy Impala with my friends and driving down the Garden State Parkway to the Jersey Shore, our perms brushing the ceiling of the car, all of us wearing Original Jams shorts and Esprit T-shirts with rolled-up sleeves. We'd load up with diet Cokes, baby oil for tanning, gum for chomping, and a boom box with cassette tapes of the Police and Prince, and off we would go to Bradley Beach in Point Pleasant, which only cost a dollar to get in. Then we'd oil up and "lay out" for eight hours, cultivating our melanomas in the broiling, polluted Jersey sun. I was the "edgy" one of my friends, because I wore Ton Sur Ton clothes and knew as much music trivia as the guys.

Jersey! I loved the bored, slutty girls who worked at the delis that their families owned, oblivious of the leers of all the men in line as they silently, sullenly handed you your change with your Taylor ham and cheese on a hard roll ("Taylor," as far as I could tell, meant salty, dark, and leathery). I loved going to Bruce Springsteen concerts at the Meadowlands Arena and yelling "Broooooce!" When I barreled down the New Jersey Turnpike with my car radio blaring, no seat belt, I was at my most ragingly alive.

The Paparazzi: Welcome, Friends

Celebrities may loathe the paparazzi, but not I. It's never a bad idea to involve them in your story. The presence of a gang of sweaty, shouting photographers can add a frisson of excitement and an action-movie element to an otherwise conventional profile. Although if the two of you are being chased, it might mean that they are mistaking you for a "galpal," which is incredibly insulting to your celebrity. They are used to scoring models and hot bartenders, not pale, spongy journalists, so do not be insulted if they make an extra, even frantic, effort to evade pursuers.

In one case, when I was in Los Angeles with Ben Affleck, this did not happen. During our meeting, he was fragile and uncharacteristically moody, having recently broken up with Jennifer Lopez. Usually press people adore him because he's bright and quippy and delivers just the right funny, original quotes that add sparkle to a piece. On this day, however, he declined to perform and I faced my usual dilemma: While I understood completely that he didn't feel like being a dancing monkey, I needed to secure a decent interview. My patter flopped (Courtney Love had given a particularly unhinged radio interview that morning and I relayed some of the gruesome tidbits, but he didn't bite), so I asked eight questions in a row about the movie he was promoting. Nice, safe ground, and I saw his tensed shoulders relax a little.

Soon, however, I had to ask him about Jennifer, and I began to sweat. Apparently they had made a pact not to talk about the relationship. "You're not going to get anything," his rep said. My editors wanted something, anything. You can squeeze a lot of publicity out of even one sentence.

I tried the "let's work together" approach. "You know I have to ask you about Jennifer," I began as my neck started to itch. He stared at me, his mouth forming a small, ironic smile. Uh-oh. Hives were starting to erupt. This always happened when things got awkward. Why, why didn't I wear a turtleneck?

"Ask away," he said with a sharp laugh. "You can always try."

We began a long dance. He claimed that the media turned the two of them into a spectacle; I gently countered by saying that they helped the media along. He argued that they didn't court the paparazzi. They didn't pose for any magazine covers and only did one or two major interviews. They were just living their lives, he said, but the paparazzi captured their every waking moment. My hives, at this point, were in full effect. It's never pleasant to confront people, but when the person is a film star, it adds a surreal element that throws you completely off balance. I prayed he couldn't see that my neck looked like a plate of ziti. Chin down. Chin. Down.

After some tensely polite back-and-forth, he decided to prove his point to me. He grabbed his keys and suggested we go get a taco at Poquito Mas, one of his favorite Mexican joints. "Just watch," he said, smoothly piloting his black Beemer into the parking lot. "This will take three minutes. Maybe four."

Sure enough, just as I was placing my order for a veg burrito, a guy in a pink shirt appeared and began snapping photos from a van in the parking lot. Frequently, Affleck said, the parking valet tips off photographers for a couple of bucks. We sat down in the taco joint's outdoor space with our trays. "Hide the tape recorder," he said quietly. I shoved it behind his supersized soda. I always try to hide the tape recorder, anyway, in the vain hope that the celebrity is lulled into thinking it's a regular conversation. We proceeded with the interview as the guy snapped away. Because we were laughing a lot, Pink Shirt thought it was flirtatious banter and ventured closer and closer.

After we finished our meal, Affleck glanced at him and said, "Uh-oh. He's losing interest. We need to look like we're hiding something." This was becoming sort of fun. As I took my tray to the trash can, I pretended to do a double take and then squished myself unobtrusively into a corner, as I had seen celebrities do. I crossed my arms and kept my eyes down. He went bonkers. Snap snap snap snap snap. "Let's hold hands," Ben whispered.

"Too obvious," I said back.

"Well, then, I'll give you a quick hug," he said under his breath. He put his arm around me. I tried in vain to relax and assume a loving expression. "You're waaaay too stiff," he whispered in my ear, which made me laugh. We walked to the car as another photographer pulled up in an SUV and Pink Shirt, three feet away, snapping continuously as he shuffled backward. This is why there are so many shots in the tabloids of famous people looking irritated. Invariably a caption will run that says that the celebrity is frowning because they're heartbroken or fat or rehab-bound or out of work, but ten to one they were just exasperated because they literally couldn't walk forward. If you stop and pose, sometimes they will drift away. Sometimes they yell things. ("Big fan! Big fan! Over here! Can you look in the camera?") In the case of Pink Shirt, he was eerily silent, even when Affleck asked him questions.

The next day, when I returned to New York, there was a bidding war in the tabloids for the photos of Ben Affleck and his new paramour. One of them paid eleven thousand dollars for the shots. The photo that ran shows Affleck, his arm tightly around me, making me laugh so hard that I'm showing some unfortunate Seabiscuit-style choppers. We looked for all the world like carefree lovers, which neatly underscored Affleck's assertion that despite many shots of himself and Lopez making out in color-coordinated outfits, he wasn't always stoking the media frenzy. After all, he was just getting a taco, minding his own business, right? He proved his point, the photographer got paid, and I got to be Mystery Galpal for the day. Everybody won.

A similar fiasco occurred during an encounter in London with Mel Gibson. We were having lunch at the Ivy, a restaurant that was the ultimate in trendiness at the time of our chat. Earlier in the day, I had joined him at a sound studio, where he was recording dialogue for *Braveheart,* the story of William Wallace, the Scottish rebel who liberated his country from English rule in the thirteenth century. In those pre–*Passion of the Christ* days, Mel still had the ability to quicken the female pulse, and on that particular morning, he was dubbing dialogue from a love scene. The sound room's inhabitants were two schlumpy sound guys, Mel, and me, so as he said the same line over and over to his onscreen ladylove, he directed it at me, just to be goofy.

"Ah love yeh," he said in a Scottish burr, staring at me intently. "Alwehs hahv." He wasn't satisfied with the delivery—it was a pivotal moment in the film—so he did the line probably ten or twelve times, while I, God help me, fell deeply in love. I like to think that my presence enhanced his acting abilities on that day, because I stared back at him with the kind of unblinking worship that even the actress playing his medieval sweetheart couldn't have reasonably conjured up.

Then we jumped into a car to head for the Ivy, while Mel and his sturdy English driver joked around the whole way. At the restaurant, he was funny and affable and occasionally ate off of my plate. The food sharing went one step further after he ordered some spinach. As he talked, tiny pieces of spinach took to the air, gently mingling with my fish, like a garnish of chopped parsley. I didn't mind.

As we chatted away, a manager stole over and whispered discreetly into Gibson's ear. It seemed that the paparazzi had been notified and were massing outside of the Ivy. As we got up to leave, Gibson instructed me to keep my head down. "Don't smile or wave," he said. "Don't make eye contact. Just get into the car as quickly as you can."

When you see photographers jostle a celebrity, it seems exciting, but it's actually disorienting and—when a lot of them have gathered—frightening. As we attempted to get to the car, hordes of hollering photographers blocked our way, flashbulbs firing. It was chaos. Head down. No smiling. Gibson grabbed my arm and propelled me decisively forward, in a very *Braveheart* manner.

"Mel!" they yelled in the tumult. "Who's the girl?" I suddenly realized that they assumed that he was cheating on his wife, Robin, with me. Which, for a man who fended off strippers, had to be a little bit of an insult, let's face it.

"Who is she?" screamed one photographer. They surrounded me as I attempted to open the car door, one of them stomping on my foot as he got pushed. Gibson's driver quickly forced the door open and shoved me inside. My heart was jumping as though I'd had a hit of crack. How did people get used to this? They pounded on the door of the car and chased us down the street as we pulled away.

When I returned to my office in New York, there was a packet of pictures sent by a news agency waiting in our photo department labeled "Mel Gibson and an unknown woman." I am smiling. I am making eye contact. I look as if I am atop a float in a parade.

4.

If you were to look at my past jobs, you'd see there is scant evidence that my life's work would involve running from photographers. You certainly can't find any in the series of "character-building" jobs I was forced to get by my folks when I was a teenager. I was, of course, too timid to get a cool job—working at a record store or being the camp counselor all the kids looked up to. My sister Heather was far savvier, landing a coveted job at an English chocolate shop in the Short Hills Mall. "We can't sell them if they're damaged," she would say when I would come to visit. Then, with a quick blow, she'd crush some cream-filleds. "Oops," she'd say coolly, pushing the flattened blobs toward me. Heather, a sugar fiend, often got high on her own supply, and eventually the place went out of business. Coincidence? She also worked at the local cinema. The owner was some shadowy, little-seen figure from Newark, and the teenage manager was her best friend, Kerry. Heather recycled tickets to sell and pocketed the profits, and, as she did at the chocolate shop, hauled home bulging bags of candy, Sno-Caps and Reese's peanut butter cups and Junior mints.

Job interviews gave me crippling performance anxiety, so I took the kind of gigs for which you basically just have to show up. My hellish trifecta of summer posts began with Morey LaRue dry cleaners. While other kids were out mowing lawns in the sunshine, I retreated into the airless chill of Morey LaRue, with its noxious odor of industrial chemicals and b.o.-scented oxford shirts. A small bonus was that in the summer, there were fewer customers, so I mostly sat in the back of the store and read Sidney Sheldon novels, sighing loudly and throwing my book down whenever a customer ventured in. As the summer wore on, I gradually turned as gray as Morey LaRue's exhausted carpeting.

Then there was my stint at Burger King, which lasted two weeks. The manager handed me an unwashed polyester uniform that smelled like it had been deep-fried and had me start at the bottom rung, at the drink-filling station. The floor of the place was thickly coated with a layer of grease and sesame seeds. "You can't really walk," said the manager, Randy. "You, ah, have to—I would call it skating." Humiliatingly, my family came to visit me one night on the job. I had just graduated to specialty sandwiches, so they all dutifully ordered the fish fillet while I waved from the back.

Far worse was my waitressing gig at the local nursing home, which had the bonus uniform of a hairnet and white elasticized polyester pants. Three options were offered up in the joyless dining room: low salt, low sugar, and puree. At least the dining room patrons still retained their mobility, and their marbles. The next part of the job was loading up the trays on a wheeled device and heading to the Shut-In Wing, passing whatever "cheerful" holiday display was situated on a table in the hallway, such as a dusty haunted house that the administrators hauled out well before Halloween. They loved to play up holidays. Hang in there! Christmas is comin', and we'll have butter cookies from a tin that have green and red sprinkles on them!

Two shut-ins in particular liked to have a little fun with me. One cod-eyed old gent had a prosthetic arm, and when he dimly heard the squeaky wheels of my cart, he would gleefully unhinge his fake limb and let it crash to the floor. "Dropped my arm," he would wheeze, and I'd have to pick it up and fumblingly reattach it. Then there was the ancient mariner who would regularly unveil his penis with a phlegmy cackle. His slack, bluish member recalled those Discovery Channel shows in which a camera travels to the bottom of the sea to film the milky, translucent creatures waving back and forth on the ocean floor.

When I am interviewing a band and I reach a silent patch, I always ask about bad summer jobs, because the answers are usually funny—pulling the tassels off of corncobs while doing bong hits, working at a doughnut shop while doing bong hits. Recalling a summer job, as well as the name of a person's first band, will always liven up the proceedings. My favorite first-band name was supplied by Dave Pirner from Soul Asylum. As a midwestern teen, he was part of the Shitz. This is great in so many ways—the spelling, the swaggering swear word. A proper first-band name must ideally be a hideous mix of pretension, hormones, and bravado: Transcendence, Are Those Real, Odyssey of the Sheep, the Crotch. Even better is a vague sociopolitical veneer—Fecal Matter becomes Fecal Matter of the Mind, for example.

But I digress. If my summer jobs were less than ambitious, I applied the same gutless approach to my first real job. New York City being a far-off fantasy, I devised a plan in which I would work my way up the ladder with the myriad publishing opportunities available in New Jersey, and then, when I was less intimidated, try New York. So I confined my job search to the pages of the *Star-Ledger.* It was there that I found the gateway to my future. "Small but growing ad agency in Cranford, New Jersey, seeks proofreader for medical advertisements. Fast-paced, friendly work environment."

Briskly, I circled the ad. Who knew where this could lead? It offered promise (growing!) and intrigue (fast-paced!). Most important, Cranford was right next to the town where my boyfriend, Ritchie, lived, so he could meet me for lunch. He waited tables at a restaurant at the mall, but he wrote poetry, too, and it was clearly a matter of time before his talent was recognized. Then we could be a media power couple.

I aced the interview and promptly landed the job. I was an adult, in a real office! My mother helped me dress for my first day as a member of the work force. She had recently gone back to work as an account executive at a furniture supply firm, so she had all of the late-eighties working-lady gear at the ready: string ties, nude pantyhose, a plaid suit from Ann Taylor yoked with gargantuan shoulder pads, and huge gold door-knocker earrings. "You have to look polished," she said as she vigorously applied brown shadow to my eyelids. For the final touch, I "scrunched" my perm with L.A. LOOKS mousse, then bent over and sprayed my entire head with Aqua Net, then flipped my hair back up for maximum volume. With a final coating of hair gloss so that my perm achieved a Jheri-curl effect, my transformation into Rick James was complete.

Unsurprisingly, I fit right in among the employees. Our outfits were the most colorful thing about the company, which had all the hallmarks of a dystopian New Jersey office: the bag lunches in the fridge labeled "Do not take," the communal couch with greasy head marks that floated over the backrest, lacquered over the years with various hair products. I was shown my oatmeal-colored, windowless office by my boss, who, to my horror and the snickering delight of the staff, had woken up that morning with Bell's palsy, so one side of her face sagged. "It's okay," she said. "It doesn't hurt. Here's . . . ah . . . here's your desk," she said, opening up a drawer for my inspection. "This is your phone, here. And, ah, here is some copy to proofread." She handed me some mock-up ads for some sort of heart medicine that kept your valves open.

Work swiftly became unbearable. When I shuffled through the doors in the morning and beheld the cheerless industrial carpeting and fluorescent lighting, my brain congealed and my private parts dropped off. Spending eight hours in my drab veal pen plunged me into despondency. "Maybe you just need to make your office more colorful," my mother suggested at dinner. The next day, I went to the local Hallmark store on my break and bought a bright poster of some large crayons in a vivid row and hung it behind my desk. Crayons. That's how far gone I was.

A new poster, my coworkers would remark, lingering endlessly in my doorway. *Wow. Looks good. Colorful! Where did you get it? Oh, the Hallmark store? Wow. Yeah. Looks really good.*

My only solace was my lunchtime visit from my boyfriend, Ritchie, who would pull up to the office entrance in his dilapidated poet-mobile.

He was the requisite bad boy one dates immediately after college—a jaunty, chronically late ladies' man who only confessed his feelings when he was blind drunk and was, thrillingly, from a blue-collar town. Ritchie was in his midtwenties and still unapologetically lived at home—next door to his sister, who left the nest to live there with her husband. Unlike the local strivers in my town who wanted to work on Wall Street, Ritchie had a gold chain nestled in his thick pelt of chest hair. Why had I not noticed how alluring a gold chain was?

In high school, my dating record had been spotty. Being in a family of three sisters was not unlike attending an all-girls' school, so I was never entirely comfortable around boys. Every victory was tinged with pathos, starting with my first kiss. A group of my middle school friends and I were bunched in a circle on May Drive on a balmy summer night, playing Truth or Dare. This gangly gang included Spencer, the cutest guy in school, the one who resembled every dreamy guy in every early-eighties movie: sleepy blue eyes, shiny brown hair that's slightly shaggy, a deep tan, perfect-fitting dark blue Levi's corduroys. He was Matt Dillon in *Little Darlings*, Jake in *Sixteen Candles*. As a bonus, he was just a little bit obnoxious. When it was Spencer's turn to play, he was told by one of my well-meaning friends that he had a choice.

"You can either kiss her"—she pointed to me—"or you can go kiss that tree over there." He wrinkled his brow. Was he really having an internal debate? Then he *looked over at the tree*. Jesus Christ. Was he checking it out? I did a quick inventory as the seconds ticked by. My hair was feathered perfectly. Yes, I had braces, but I had carefully Water Pik–ed them before leaving the house. Calvin Klein jeans, my "good" plaid shirt (it had silver threads in it), and a generous spritz of Love's Baby Soft. After what seemed like an hour and forty-five minutes, he grudgingly picked me. That was my first kiss. I can only be grateful that he didn't think the tree was foxy.

As I went on to high school, I was the funny girl, so my only real victory was my senior year boyfriend and prom date, Mark, who had a dream résumé: impeccable musical taste, dimples, a soccer player, and he worked at the Ralph Lauren store at the mall, so he had great clothes. Being the Phyllis Diller of any group when I yearned to be the Ava Gardner, I was always a sucker for a carelessly handsome rake, and when Mark cheated on me with a junior girl, it was horribly thrilling. The drama of his transgression made my senior year one of the happiest of my life. But there was that lingering feeling that he was ever so slightly doing me a favor, so I was always on my best behavior. This would eventually serve me well when it came to celebrities, because I became adept at snappy material, anecdotes with never-fail punch lines, choosing the perfect outfit. Because I was constantly performing, Mark certainly didn't get a glimpse of the real me, which I assumed would bore him, anyway. But I did the same thing back to Mark: I didn't necessarily go out with him because of who he was, but rather what he represented.

It was the same thing with Ritchie. He was naughty, and unpredictable, and would sometimes show up three hours late, or not at all. His life had no structure, while mine was carefully regimented. From birth, my folks had imposed upon us a rigid schedule of chores, lessons, and homework. Our house was always in perfect order, and I had absorbed their compulsively neat ways so thoroughly that I took great satisfaction in putting the groceries away with the can labels carefully facing forward. I was dazzled by the fact that Ritchie simply didn't care that he had no Life Plan, or even a weekend plan.

After I spent a few months with Ritchie, his big, hard-drinking Irish family took me in, and I proudly joined in on their yearly pilgrimage to the St. Patrick's Day parade in New York, where we would descend on the Midtown Irish bars, all of us wearin' o' the green sweatshirts (his mom with a green Tam o' Shanter), and get boisterously drunk by lunchtime. On more than one occasion, I contributed my own tributary to the emerald rivers of barf that flowed through the streets like the Shannon.

Nothing gave me more of a charge than to watch Ritchie's three brothers on a Saturday as they tinkered with a car engine for hours. My dad, meanwhile, had "a guy" for automobile mishaps.

"Jesus H.," said my mother, who got especially southern when she was worked up. "How can y'all not see what a loser Ritchie is?"

"You're just a snob," I would shrill back. "It just bothers you that his family parties together."

"It has nothing to do with money," she said sourly, with a firm jut of her lower lip. "A loser is a loser."

Well, he was the small-town king of his high school friends, all of whom admiringly talked about Ritchie's likeness to Jim Morrison. "And he's a poet, just like Morrison was," they'd point out, watching Ritchie moodily sip his tenth Molson of the evening. Ritchie would take me to noisy parties, where his freewheeling pals seemed to have more fun than my more conservative friends. Everyone at my school was too uptight to have costume parties, or softball games sponsored by Dick's Auto, or go camping in the Pocono Mountains with five cases of beer, or dance to bad R & B like Klymaxx or L'Trimm. *I can fit in anywhere,* I would think as we formed a cheering ring around Matt's brother-in-law's beer bong. "Smell ya later!" I would cry after a night of revelry.

"Take ceh!" they would chorus back.

At one of these shindigs—it might have been a Halloween party where, inevitably, there would be one white guy who would dress up as a black man with an Afro wig—I met a girl named Amy who worked in the marketing department of *Rolling Stone.* She was doing the usual post-college activity—hanging on to high school buddies whom you will eventually stop calling. As the celebrants around us started up a game of Beer Pong, I quizzed her about her job, which seemed unimaginably glamorous. I'd subscribed to *Rolling Stone* for years, and indulged in a ritual when I read each new issue: start with the Charts page in the back, then read "Random Notes," the gossipy column in the front. Then move to the reviews and, finally, to the cover story.

I remembered as a teenager hurrying with Heather to catch the 1985 John Travolta movie, *Perfect,* the story of a *Rolling Stone* reporter, on opening weekend. The film commenced with Travolta as a restless reporter who was plenty tired of working the obits desk of the *Jersey Journal.* He wanted more. At the time, I was an intern at *New Jersey Monthly* magazine, fact-checking a column called "Exit Ramp."

"He's just like me," I whispered to Heather.

Cut to Travolta a few years later in New York City, working as a reporter at *Rolling Stone.* I watched, openmouthed, as he interrupted his fast-paced life long enough to lunch with his boss, "Mark Ross," played by Jann Wenner. Soon they started kicking around story ideas, because his mind was always going-going-going.

Travolta ogled some women who walked by in full-on eighties aerobics regalia: candy-pink butt-floss bodysuits, puffy Olivia Newton-John "Let's Get Physical" headbands. "Why not do a story about how health clubs are turning into the singles bars of the eighties?" he says. How, I marveled, was he able to just pluck that idea out of the air?

"Mark" mulls the pitch over for about a nanosecond. "Hot tubs? Alfalfa sprouts?" he muses. "We haven't done California in a long time."

Next, we see Travolta at a Los Angeles health club, leering at "Slimnastics" instructor Jamie Lee Curtis. After she rebuffs him a few times, he finally wrangles a meal with her, but is unable to persuade her to be interviewed. After lunch, he is driving back to the Sunset Marquis hotel and pulls out a tape recorder. "Notes on lunch," he announces briskly. "She's smart, but I've gotta be smarter."

Later, he is introduced to a guy at the gym. "I loved your Carly Simon piece," gushes the man.

"So did I," John modestly replies.

I was mesmerized, utterly blind to the cheesy dialogue and hackneyed plot (no reporter in their right mind talks into a tape recorder, especially to say "Notes on lunch").

"You could easily do that job," whispered Heather, ever steadfast.

"Right," I said. Surely I was just the kind of person *Rolling Stone* was searching for: a Jersey girl who pulled down B's and C's in a state school, with long hooker-red nails and a passable knowledge of music.

Meeting Amy was my chance. Emboldened by my seventh Old Milwaukee and inspired by the success story of a fellow Garden State girl, I asked if I could send her my résumé.

"Sure," she said, scribbling down her address.

Ritchie spotted us chatting and wended his way over through a gauntlet of high-fives.

"What's crappenin'?" he said, slinging his arms around both of us.

"I'm bothering your friend here about a job at *Rolling Stone,*" I said, flashing her what I hoped was a warm, conspiratorial smile.

Ritchie squeezed Amy's shoulders more tightly. "She doesn't need to work at *Rolling Stone,*" he said. "My girl is doing just fine right where she is." He gave me a loud kiss on the cheek, the kind that goes on for a full minute until everyone around you is uncomfortable.

I smiled at her. *Ignore him,* I telegraphed.

Ritchie's brother-in-law Raymond, still in his green maintenance-worker uniform, barreled over and pulled me down next to him on a beer-stained plaid couch. He pointed at the window. "Look over there," he said loudly. "It's a bunch of seagulls flapping at the window, trying to get in. You should probably close your legs."

"Raymond!" I hollered, punching him. I was so loose, so free!

"I'm just jokin' with ya," he said, enveloping me in a b.o.-scented bear hug while Amy looked on.

I figured Amy would chuck my résumé in the trash, but a few weeks later, I was flabbergasted to receive a call from someone in the magazine's editorial department, asking if I was interested in interviewing for a job as an editorial assistant. It wasn't very dignified to have my mom answer the phone, but I put on my best professional, slightly nonchalant voice.

"Tuesday?" I said. "Let's see . . ." I pretended to check a date book while my mother hovered.

"What?" she hissed. "What are they saying?"

"Ah," I said smoothly. "Yes, Tuesday works."

How to Control the Panic When Your Subject Is Absurdly Famous

Why, you may well ask, do I get so nervous before interviews? They are, after all, just people, right? Mostly, it's money-related jitters: If celebrities do not supply the amusing quotes that the story requires, the story could be killed and I will not get paid. There is a realistic danger that, during the measly forty-five minutes I am allotted for our chat, a famous person will take it into their head to natter about some subject that I cannot possibly use in the story: where they were on September 11, for instance. For years afterward, famous people would steer the conversation toward the Tragic Event, and I would think, *Oh no. Please, no. I can't use this, and everybody in America has a September 11 story, and it's very hard to interrupt you people when you get rolling. Just give me the funny on-set anecdote and let's call it a day.* There are so many conversational sand traps that could eat up a precious fifteen minutes, from new-age philosophies to complaining about paparazzi to the hands-down most dreaded topic among interviewers, My Craft. It always placates actors if you toss them a throwaway question or two about their acting process, but as it is hard to get them to stop, it is best to avoid the subject altogether.

Then there is the familiarity factor. When someone who is weirdly familiar is cranky, it's disconcerting. I am a shy person and acutely aware of the barest flicker of moods in others. I wish I could be one of those Teflon interviewers who can obliviously stick a mic in someone's face without noticing their defensiveness, but I can't. I can only relax once they do.

Sometimes, if the star is of particularly high wattage, I need a little extra help. That's when I pop a brand of pills from the health food store called Calms. They basically quiet the screaming in my head without a buzzy, druggy feeling. Who knows, maybe they're placebos, but I find that my hands aren't quite as Niagara Falls–like if I've gulped down two or three Calms. I recommend them to tamp down anxiety in all but the most hard-core instances. Which, in my case, was my interview with Madonna.

When I heard I was going to have a sit-down with arguably the most famous person in the world, you can best believe I hotfooted to the health food store and bought two bottles of Calms, particularly after one of the people at her record company advised me not to act afraid, because she smells fear, like a dog.

I could understand perfectly why she wouldn't want to deal with people who quake in terror. It must get tedious, joking that you "won't bite." But while I could empathize, I was still paralyzed with terror when I traveled to her Maverick Records office in Los Angeles. Most women in their twenties and thirties who have grown up with her have a proprietary relationship with her that transcends fandom. My friends and I have maintained this connection even as our interest in other stars has quickly faded. To this day, I will read any item on her and study any photo with the zeal of a Talmudic scholar.

As I headed over in a tele-car, I realized with alarm that the Calms had not kicked in at all. From here on in, I vowed, I would get a prescription for a nice tranquilizer. I gulped my fourth Calms, which lodged in my throat. Goddamn Calms! Maybe I just needed to build up some residue in my system and then they would work.

My hands, as usual, were sweating. Lord, what if she shakes my hand? I rapidly clapped my hands in order to dry them off, as the driver glanced sharply at me in the rearview. Why didn't I bring a tissue? I inspected my palms in the lurid California sunshine. They were glistening.

The driver stomped on his brakes. "Son of a bitchin' bast'!" he screamed. What was that dialect? And what was he distressed about? There were no other cars on the road. I looked for a squirrel, or a bird. Nothing. He was not helping my nerves.

I leaned forward and asked the driver for a tissue. "I don't haf, miss," he said.

"Okay," I said. "Thanks, anyway."

"I don't haf." Colombia? Russia? Iceland?

"Right. Well, thanks."

"Glove deportment. But I don't. I'm still the King."

"Right." Now my hands were sweating in earnest, because we didn't have a language barrier but a sanity barrier. God, if only this were a restaurant interview! Then I would use one of my little hand-drying tips, which I pass on to you: Get there first, order a cold drink, and clutch it in your palms so that they stay cool. Use the condensation on the drink as a hand bath, and then, when it's time for the Big Shake, swipe your hand on a cloth napkin as a sort of abbreviated wash and dry. Maybe I could request a can of soda or something from an assistant. Or maybe Madonna wouldn't shake hands. A lot of famous people are germophobes. Again, who can blame them? All those clammy hands that you must shake in a meet-and-greet, encrusted with God knows what? Yecch.

We passed a parks employee, desultorily sweeping the sidewalk. *I wish I were you,* I thought fervently. *Sweep, sweep, sweep. Aaand repeat. Why can't I be you?* (This had also been directed at my cat, curled up in a tranquil ball as I left the house for the airport.) We traveled onward, passing a taco hut. Its patrons stared blankly out the plate glass window. *I'd rather be you, or you, or you. Even you, with the port-wine stain.* In three hours it will all be over. In three hours it will all be over. Please, Jesus, let her be in a good mood. Some people love being driven around in a limo, but the only time I ever take them are on the tense, gloomy journeys to interviews, so to me, they're the Transportation of the Damned. Well, with one exception. After I finished up a dinner interview with Melissa Etheridge, easily the nicest person I have ever profiled, she had her driver drop her off at her Los Angeles home before continuing on to my hotel. As she said her good-byes, she told me that she had paid for the limo for the rest of the evening, so if I wanted to go somewhere, I should feel free.

"I'm starving," I admitted to the driver, a large, easygoing guy named Rodney.

He laughed. "Didn't you just eat?"

"I never really eat at interviews," I said. "Are you hungry, by chance?"

"I am, actually," he said.

"I could go for a cheeseburger. Or maybe some barbecue." After an interview, I always liked to treat myself to something hi-cal. I had planned to get an ice-cream sundae from room service.

"I know of a great barbecue place," he said, brightening. "But it's not in a part of town that you would consider nice."

"I'll treat if you take me," I replied. Off we went, Rodney and I. He was amazed to see how much pulled pork I could put away.

I was jolted back to the present tense when the driver stomped the brakes once again. Jitters. Jitters. After the fretting part, the self-chastisement. Christ, it's not a medical procedure. It's just a profile. How about some perspective? And how about not praying to Jesus with a request to put some celebrity in a good mood, when clearly He has more pressing problems? Plus, remember your pledge that you would only pray to thank Him for your good health and wonderful life, so as to get on His good side, and save up a prayer coupon for only the big stuff.

This is inevitably followed by an out-of-body feeling. How on earth did I get this job? Clearly, I'm not at all qualified. But who is? What are the qualifications? I'm from Jersey, for Christ's sake. Although isn't everyone from Jersey? Is Long Island really that different? Is Philly? The driver stomped the brakes again. "You're foot-sick!" he screamed to no one.

As the car finally rolled up to Madonna's office, I had the familiar panicky feeling of wanting to leap out and just sprint down the street. Well, what if I did? What would happen? Would the world end? No, it would not. I looked at my watch. Oh, Lord, we're twenty minutes early. *Don't be late,* the record company person told me. Apparently Madonna, ever the professional, did not tolerate tardiness and had even canceled interviews because of it.

I had to go to the bathroom. My hands! Ugh, they were like soft, moist frogs. I had to do something—otherwise she would shake my hand, be repulsed, and the interview would be over before it started. The more I concentrated on them, the wetter they would be. For the love of Pete, think of something else! Something! Else!

I wondered if I should wait in the car. Yes. It's better than hanging around some lobby, pretending to examine the prints on the wall. Plus, I might faint. She was just too famous. A sitcom star, that I could handle. This was something else entirely. "Sir?" I said. "I'm just going to wait here for a few minutes, if you don't mind."

One staring eye, meshed with veins, was visible in the rearview mirror. Why wouldn't he turn around?

"On second thought, I'm going in." I walked into the lobby, which had a large photo of Madonna's eyes on the wall. All the men in the office were sleekly trendy, the women less so. The receptionist told me to take a seat. In my head I went over my questions, which I had memorized. If you consult a list of questions, it tends to break the momentum and your chat will be less conversational. You want to at least create the illusion that you are simply two friends or associates having a nice little confab. Although, as I reminded myself, I had to remember to start off with questions about her album. Always, always lead with queries about the person's project, the reason why they are granting the interview. If the questions are not presented reassuringly up front, your subject will get visibly riled.

An assistant broke my trance. "Hiiiii," she said. "So Madonna's ready for you." I couldn't breathe, but I followed her down the hallway, surreptitiously wiping my hand on my sleeve. If only it were the fifties and I could wear gloves. What else did the publicist say? *No chitchat. Get right down to business.* Which I can respect. Why bother with the blah-blah? We both had a job to do.

We were in her office and oh Christ, there she was. I had probably seen her face more often than I have seen my own. Smells fear! Like a dog, smells fear! Dog! Fear! Smelling fear! Run away run away run away!

I shook her hand and looked her in the eye. At that point, I was so filled with terror that my body reversed its natural inclinations and my palms were bone dry. "Hello," she said. Cordial, but all business. As usual with every single famous person except Clint Eastwood and Uma Thurman, I found that she was smaller in person: five foot two. And, at the time, pregnant with little Rocco. It was odd to see her heavily pregnant, because despite being one of the globe's most photographed women, hardly any photos existed of her with child, assumedly not a coincidence. Because she didn't look like the image I had of her, I was almost able to pretend she was someone else. This quieted the internal screaming, somewhat.

Time to show my lack of fear. Deep breath! "I just read an interview in which you complained about your adult acne," I said, scanning her face. "What a load of shit! That's just something you say to make us feel better about ourselves." Sassy, yet unctuous! I saw her smile a little. Good. Then I pulled out my I Get You question, about a book that I remembered she wanted to option for a film years ago, Jeanette Winterson's *The Passion,* which I, too, had loved. What came of it? I asked, and she sprang to life, telling me that she once wrote a letter to the author and never heard back, and how disappointed she was.

Then I started right in with questions about her new album as she slowly, gingerly lowered her pregnant body into a chair. It made her uncharacteristically vulnerable, and slightly less frightening than she would have been in her *Sex* book days. At one point, a crazy montage of all of her videos, films, and life events flashed through my mind, unbidden—*Remember that video when she was a redhead? What was that? "Fever," right*—and with a Herculean effort, I tamped it down. Four questions down. Good. Good. Then it happened: As she talked away, I realized that I couldn't remember my next query. My notebook was in the car in the car in the car, and when she finished talking about how her spiritual quest informed the album, we were just going to stare at each other in hideous silence.

Take it easy, I counseled myself. Pull out some lightweight emergency questions that you usually reserve for interviewees who keep checking their watch. Usually it is the first time they have heard these questions, so they are fun for your subject to answer, and you can avoid the dreaded prerehearsed response. No one wants to answer for the three thousandth time the question about her musical influences, or how this album is different from the last one. Instead: What did you think about before falling asleep last night? What day did you see your parents differently? What smells remind you of childhood? What can't your friends tease you about? When is the last time that you were truly content?

I threw a reliable one at her (what was your worst high school job?) to gain some time until the other questions finally reappeared in my head. She answered immediately that it was cleaning houses, and it was gross, and she had to clean the toilet bowls of boys she went to school with. Then I shored things up with a few more album questions. As the end of our chat loomed all too quickly, I peppered her with the kind of regular old lifestyle questions that I, and all of my friends, were curious about: What was the last movie you rented? (At the time, it was Ice Cube's *Last Friday* because it was her husband's, Guy's, turn to pick.) Do you ever cook at home? No, but she helped out Guy by adding "accessories" to the salad.

"Accessories?" I said. "What, like a belt and gloves?" Look—she's laughing!

We moved on to her pregnancy. She said mournfully that her doctor informed her that she couldn't exercise, and she couldn't go out or wear cool clothes or go dancing and she just felt like "a domesticated cow." I nodded sympathetically, pretending I was her girlfriend and she was confiding in me, like she did with Sandra Bernhard in *Truth or Dare,* when she was saying that there was no one left to meet because she had met everybody.

I leaned forward and asked her something that I had always wondered about: Did she ever feel insecure? She rarely exposed any sort of vulnerability. "I feel insecure every five minutes," she shot back. "What are you talking about?" She said that being as pregnant as she was, she panicked when she looked in the mirror.

I pressed her, because I wanted to know how she felt when she wasn't pregnant. Say she came across a picture of a boyfriend's ex. Does she make mean comments? She said that there was a whole process that happens. "First I go, 'Oh, she's skinny and pretty.' " She grinned. "Then I think, 'Oh, but I'm *me*.' "

God love her! There was a soft knock at the door. The publicist. I remembered that there was something else I was supposed to do. My friend Susan, a fashion marketer, wanted me to inventory Madonna's bathroom and report back. I couldn't let her down. Plus, I wanted to know what was in there, too. I scanned her office. No bathroom. It must be right outside.

A knock, again. I took my leave. "Thank-you-so-much," I said, swiftly gathering my things and returning Madonna's firm handshake. Keep it brief. Don't smile, don't babble. And no pictures or autographs—as a professional courtesy, you're never supposed to ask. You want to at least fake that you're contemporaries.

"May I use the facilities?" I asked the assistant, pointing to the door near the star's office that said "WC." I raced in and started running the water, while taking inventory of the bathroom. A bottle of Fracas perfume, I scribbled down. Some sort of face spray that you get at the health food store, water with a geranium scent by Tree of Life. Bathroom reading? *The Hypochondriac's Handbook*. Hmm. Interesting. La Mer face lotion. Done, done, and done.

The assistant awaited to walk me to reception. I strode through the halls, triumphant. "How was she? Isn't she amazing?" she said.

"How was she?" said the receptionist. "Totally great, right?"

"She was," I said, trying and failing not to sound like a deranged fan. "She was funny, but she had a softer side, too. And she never gave canned answers, she really thought about things." Then my knees started to buckle. "Can I sit for a second?" I said. "I feel a little faint."

The receptionist nodded. "That happens sometimes," he said. "I think I'm going to get some smelling salts and put them behind the desk."

As I stumbled to the car, the cycle was complete. It always ended with me in a victory march, thinking, *I have the world's best job.* This euphoria lasted precisely as long as it took to write the story. And it was alarming to know that the Calms worked perfectly well on most stars, but not the triple-A list.

The driver eyed me in the rearview mirror. "You want to take monan, or fleen?" he said.

I pretended to deliberate for a moment. "I guess let's try fleen," I said.

5.

On the night before my job interview at *Rolling Stone,* I read up on music as much as I could, and studied the backs of my album covers. I always figured that in dealing with the unknown, I couldn't be too prepared. I already knew the magazine's history: founded by Jann Wenner, his future wife, Jane, and critic Ralph S. Gleason in October of 1967 in San Francisco, it was to be the *Paris Review* of music magazines, with lengthy, intelligent articles on musicians and the music they made. I always noticed the magazine's bylines, and all of the authors were like old friends to me: Ben Fong-Torres, Charles M. Young, David Fricke.

I gave Heather some back issues and had her quiz me on music trivia. Somehow, in our naïveté, we presumed that the editors would grill me on my musical knowledge. Maybe there would be a pop quiz of some kind. Best be prepared. "Who produced the new Pixies album?" she asked, studying my *Doolittle* record.

"Gil Norton."

"Good," she said. "They're going to hire you on the spot." If everyone had a Heather, there would be no wars. Shy at school and cheerfully, relentlessly talkative at home, Heather was so unfailingly upbeat that many a night after I rolled home from a high school keg party, I felt perfectly comfortable rousing her out of a sound sleep to make my friends and me some food.

"Heath," I would say beerily, clumsily shaking her awake. "Can you cook us some of your famous French fries?" (They weren't "famous," she just dropped them in oil.)

She'd blink dazedly, and then smile. "Sure," she'd rasp, reaching for a robe.

Before we were able to drive, Dinah and I would coerce her into riding her bike for two miles on the highway to Shop Rite to buy us snacks. "What if a guy from school sees me?" she fretted. "It's embarrassing to be seen with two shopping bags stuffed with Entenmann's chocolate chip cookies hanging off your handlebars."

"Time's a-wasting," said Dinah, handing her ten dollars. "You can keep the change."

Heather was fiercely sentimental but would skitter away from any overt displays of actual emotion, preferring, instead, to write it in a card. *I love you more than anything* was a typical message, but if you looked up in happy surprise, she was stone-faced, as if it were some sort of hoax. She was less gullible and more conniving than Dinah, and often got her way by quietly, sweetly steamrolling over you, and you wouldn't know it until you were flattened.

After Heather helped me prepare for the interview, I lay in bed all night, staring unblinkingly at the ceiling, until it was time to tremblingly get dressed in my mom's trusty plaid Ann Taylor suit, fighting back nausea. I took a bus to the city and, afraid I would be late, hailed my first cab. As I was stepping out of the taxi in Midtown, unaware of the city custom of checking to see if anyone was around before opening the door, I heard a thud and a crash. I had rammed the door right into a speeding bike messenger, and his nose was gushing blood. "You fucking bitch," he spat, clutching his nose.

I stared at him, numbly. "I'm so sorry," I faltered.

"I'm so sorry," he squeaked back, imitating me. I stood, rooted, as he climbed back on his bike, wiping the blood with the back of his hand. "Sorry," I burbled again, and then turned toward the office.

To my horror, he followed me, pedaling furiously on his bike. "You ugly bitch," he shouted as a passerby turned and gawked. "You ugly, fucking bitch." "Bitch" rolled off my back, but "ugly"? He knew how to hurt a gal. As my face burned, he followed me for three blocks, swearing at me all the way, until I fled into the office lobby.

I was still an hour early. "Sorry, man, no one's in yet," said the receptionist, a gaunt guy with scraggly long hair and chunky black glasses. "Waaay too early." I took a seat and discreetly watched as the employees straggled in.

They were the hippest people my suburban eyes had ever beheld. Most of the males cultivated a slightly grubby look. One stubbly guy in a Clash T-shirt and army pants slumped by with a cup of coffee, trailing the odor of stale cigarette smoke, probably from some show he had been to the night before. My eyes hungrily crawled over everyone. The girls, I noted with dismay, all had sleek, shiny hair. No perms? And they barely wore any makeup. No brown eye shadow? I had applied it with a trowel that morning in order to look "professional." Another girl wafted in and I studied her look: Levi's, a T-shirt, some sort of rugged brown hand-tooled belt from India, and expensive cowboy boots. Where do you buy a belt like that? Do you go to India?

Everyone on staff was probably well traveled, while I had left the country exactly once. Dinah had gutsily spent a semester in London, and my mother and I went to visit her. On the plane ride over—my first transatlantic journey—a harried flight attendant made an announcement. "Is there a doctor on the aircraft?" she asked. "If so, please press the call button." A long silence followed, and then a hasty conference with another flight attendant. "Is there a nurse on the aircraft? A nurse?" Nothing. Right about the time that it devolved into a plea for a veterinarian, my mother jabbed her call button.

"Excuse me, miss," she said crisply. She was always scrupulously polite. "What seems to be the problem here?"

The stewardess looked abashed. "I'm afraid someone has had a heart attack in first class," she said.

My mother raised an eyebrow. "Well, is he . . . is he . . ." I thought, *Oh, please don't say a goner. That's what you're going to say.*

"It looks that way," the stewardess said.

"Well, then what happens?" she demanded.

The stewardess lowered her voice. "I don't want to alarm the other passengers," she said. "In this type of, ah, unfortunate incident, our policy is to put a blanket over the person to make it seem like he is sleeping. Then the body is removed after the flight." My mother nodded. It made sense. What are you supposed to do, stuff him in the overhead bin?

Another sophisticate glided in, the height of eighties chic: white T-shirt with a gray men's vest over it, black leggings, a short black skirt over the black leggings, and Doc Martens. How did she know how to wear the skirt over the leggings? Was there some sort of demonstration at a store, or perhaps a seminar in the East Village that you could attend? And why wasn't she wearing any color? According to my mother's *Color Me Beautiful* book, I was a Winter, which meant that I should dress in "strong, vivid shades such as red, royal blue, and teal," so I duly put on a royal blue blouse underneath my plaid suit. Another girl wore combat boots and a loose dress, her hair pinned up artfully. No scrunchy? I looked down at my gloves, which I'd bought at a local store when I landed the interview. They were black, with long black fringe circling the wrist. I thought they would be hip. I fingered my giant Salt-N-Pepa gold earrings, which had been a big hit at the ad agency. All of the girls' earrings were tiny and silver and looked exotically ethnic, like they had been picked up on a backpacking trip through Guatemala.

I was trying to scrub off some of my makeup with the tissues that my mother had tucked into my purse when the door opened and a smiling woman in jeans called me in. Self-consciously, I wobbled over in my high heels. Next to her, I looked like a Mary Kay representative. Walking into the chaotic office wreaked havoc on my few remaining nerves. It resembled a college dorm, with a rabbit warren of offices, all of which had a stereo, with different music blaring from each one—My Bloody Valentine, Fine Young Cannibals, Living Color. "I'm the Cult of Personal-i-teee," roared out of someone's speakers as a couple of people stood around. Apparently it took everyone a while to get rolling, despite the fact that the office opened at the reasonable hour of ten a.m.

"These lyrics are stupid," said one guy, with a diffident wave. " 'Like Mussolini and Kennedy, I'm the cult of personality'?" He rolled his eyes. " 'Like Joseph Stalin, and Gandhi.' I don't think even he knows what he's saying."

A pang of pungent envy pierced my gut with an almost physical force. I desperately wanted to stand around with those people and dismiss bands. I, too, wanted to slurp coffee from a blue paper cup from a Greek diner that said We Are Happy to Serve You and wear jeans to work and go right from the office to dinner and drinks, then a show, then another show. Oh, how I longed for someone to ask me where I worked, so I could answer *Rolling Stone* with a falsely modest questioning intonation, as if it were some obscure trade magazine.

We passed an office that was jammed full of editors who were having some sort of ideas meeting. Some of them were shouting with laughter and clapping. Those quips were zinging back and forth! Prior to this, the closest I had ever come to a "creative" office was seeing episodes of *thirtysomething*, in which Michael and Elliot shot Nerf hoops in their ad agency office to start the flow of ideas. I stared, hungrily.

With a kind of mute hopelessness, I knew that I didn't have the barest chance of getting this job. Before I walked into the office, I actually thought I had a fix on New York life because I read *Spy* and the *Village Voice*.

As I followed the woman to an office, I saw that every desk was piled high with dozens, sometimes hundreds, of advance cassette tapes (remember, it was 1989) and all manner of intriguing packages from record companies.

An assistant, self-assured but offering a friendly smile, waited at a table with a notebook. She had four earrings on the top of her left ear and a long braid down her back. "Hey," she said. I started to stick out my moist hand but retracted it when I saw that she wasn't going to stand up. "Don't be nervous," she said. "I just have a few general questions about what you like to do." She laughed. "It's better than talking about your grade-point average, right?" Lifestyle questions? Was she asking me lifestyle questions? Dread bubbled up in me like a geyser.

"So, uh . . ." She consulted her notebook. "Hm. What do you like to do on weekends?"

I ran through all of my recent activities: a party at my high school friend Frank's house "down the shore," a trip to the mall, a party at Ritchie's cousin's place where they gave the dog beer. That was my life. I trawled my brain for something that was even vaguely cool. "I go to Maxwell's a lot," I said. "I guess the most recent show was Robyn Hitchcock." Maxwell's was a tiny club in Hoboken that I used to frequent. Hoboken, a commuter town that was teeming with twentyish Jerseyites, was as far as I ventured from home.

She looked up from her notebook and brightened. "I was there, too!" she said. "Good show, right?" Thank you, Jesus. I rummaged my brain for more hip activities. I thought of my sole trip to London, two years ago.

"I love to travel," I blurted.

"Oh?" She scribbled something down. "Where have you been lately?"

"Lately?" I pretended to reflect. "Well, probably my most recent trip was to London."

"Cool," she said. "What did you do?" Well, let's see. I saw Big Ben. I rode around in a red double-decker bus with my mom alongside some nice people from Pittsburgh. I had high tea at Harrods. I watched the changing of the guard. ("Mom! Take my picture in front of this guy with the red coat! Get this, he's not supposed to change expression!") Oh, my God.

I shrugged. "Oh, you know, the usual. Went to clubs." Please don't ask where.

She nodded, smiling. "What were the last five records you listened to?" Time to lie! I fancied myself something of a music connoisseur, but the bald truth was that my record collection abounded with clunkers, mostly Jersey dirt-ball rock such as Van Halen and, thanks to Ritchie, an inordinate amount of bad late-eighties R & B. Oran "Juice" Jones in the house! Thinking quickly, I selected one record from each category, starting with Venerated Jazz: Miles Davis's *Kind of Blue.* Rock? The Clash's *Combat Rock,* and that was actually true. Maybe something a little trendy. Neneh Cherry, *Raw Like Sushi.* She nodded again. Good. And, because I was going through the requisite reggae phase endemic to suburban white kids in the eighties, how about we throw in Bob Marley's *Babylon by Bus,* the live double album. A live album would nicely convey my love of music. Sweat was pooling in my bra as I contemplated the last selection. I needed something that sounded authentically random, so she wouldn't think I was filling quotas. *Cypress/Afoot* by Let's Active, which I'd played that morning, until my father told me to turn it down.

She scribbled away. "Okay. Is there anything else that you do for fun?" Carefully avoiding the actual answer, I invented a sophisticated schedule—concerts, museums, and, most egregiously, "jogging."

She stood up. That was it? "Time for you to meet Bob Love, the editor," she announced. I was ushered on weak legs into his office. Bob was crisply courteous and nattily dressed in a beautifully cut dark suit. I settled stiffly into a chair and gave his bookshelves a covert scan for clues on how to behave. Harold Brodkey, John O'Hara, Sigmund Freud: brawnier fare than the Victorian drawing-room novels that I went for, so a chat about literature was out. A guitar was propped in the corner. "Oh," I could say. "You play?" Then he would say yes. Then I would have no follow-up question because I do not play, and we would sit in silence. As he sat down at his desk, I eyed his colorful socks, which were yellow and red and blue in some sort of dot pattern. Aha. A clue. *Gay,* I thought knowingly.

"I see you looking at my bookshelf," he said. "Who is your favorite author?"

"I'd have to say Truman Capote," I said, figuring I'd score some gay pandering points with Bob, who was in reality vigorously hetero and dating a string of women.

"How about your favorite *Rolling Stone* writer?"

That was an easy one. "It's a tie between Kurt Loder and Cameron Crowe," I said. I told him that for the past eight years, a poster had hung above my bed of a Bob Marley *Rolling Stone* cover, his arms outstretched, a big grin on his face. I loved that cardboard poster so much that I never took the plastic wrap off it, which made it look like one of those plastic-covered couches at your grandma's house.

After a few more of the usual queries about grades and the like, he finished up with one last question. "Why do you want to work here?" he said, leaning back in his chair.

How could I answer? Miserably, I stared at him. I wanted to tell him that I wished I had never had a hit of this particular crack pipe, so that I could live blissfully in New Jersey and never know what I was missing.

"Well," I said haltingly. "I know I didn't go to an Ivy League school—"

"Neither did I," he put in, smiling.

"And my résumé isn't exactly spectacular." He didn't contradict me. "But I want this job very badly," I continued, trying to keep the pleading note out of my voice. I stood up. "Look," I said, pointing to my résumé on his desk. "You can see that I can work hard. I used to deliver pureed food to senior citizens, wearing a hairnet and polyester pants. And look"—I pointed to *New Jersey Monthly*—"I have fact-checking experience." I pasted a ghastly, can-do smile on my face. "You have facts, I can check them."

Give me the job. I want it more than the others do. I am not "doing the rounds," interviewing at other magazines. I have no connections. This is it. For Christ's sake, give me the job, because I have never had anything even vaguely interesting happen to me in my entire life. The only distinction I have ever received was being named Class Clown in high school. I know that I don't belong here. I saw some of the girls exchange looks when they saw my perm. I am average in every way. But I know just enough to be aware that I am average.

A group of assistants was clowning around outside of Bob's door. "Who wants to order lunch?" one of them said absently. *I do. Oh, I do. I want to order lunch.*

Bob shook my sopping hand and I fled to the Port Authority bus station, home to Jersey, home to my house in the suburbs.

"How was it?" my mother asked when I burst through the door. She was making hamburger stroganoff, one of her southern classics. The recipe is simple: Brown some hamburger, then open up a can of cream of mushroom soup and dump it in. Then add a can of cheese soup (when it is thwapped into the skillet, it will retain a jiggly can shape); finally, pour in a can of water. Stir until gray and mildly lumpy. Serve over a pile of white bread that has been ripped into bite-sized chunks. "Aaah," I said, inhaling a steamy, chemical whiff. My mother stopped cooking almost entirely when she became liberated in the early eighties, but this was a special occasion.

"Jay!" my mother hollered upstairs. "She's home!" At the time, my sisters were away at college, so it was the three of us.

"Hey, kid," said my father, bounding down the stairs into the kitchen for his nightly scotch. "Did you knock 'em dead?" I felt the tears forming at his good cheer.

"Not exactly," I said as my mother handed me some slices of white bread to rip into pieces. They both exchanged looks. As we sat down to eat, I told them about the whole thrillingly horrible day, while they nodded, concerned.

"Did you tell them that you hadn't graduated yet?" my father put in. "Remember what we talked about? That you're going to summer school, and you'll have your diploma by next year?"

I forked up some hamburger stroganoff. "Not exactly," I said.

"You didn't tell the truth?" he said. I could see him begin to fibrillate. My father, in many ways similar to any fifties-era TV patriarch, always, always told the truth.

"Look, Dad, there's no way I got the job, so it doesn't matter, anyway," I said wretchedly as the tears started in earnest. "And what's worse is, I can't blame it on a mediocre state school or so-so grades," I added, sniffling. "They asked me all this stuff about my lifestyle, so when I don't get the job, it will be because they didn't like me, the person."

"Well," my mother finally said. "Since you doctored up most of your answers, then it's really a fictitious person, anyway."

"I'm sure you knocked them dead," my dad said in a hearty voice, fastening a cheery smile onto his face. Then we ate the rest of the meal in silence.

A few days later, I returned wearily home from the ad agency.

My mom looked up from the kitchen table, where she was sorting mail. "How was work?" she asked.

I sighed. Really, where to begin? There was yet another birthday celebration in the conference room, this time for Shauna in Accounting, and as I ate my piece of supermarket cake frosted with bright blue icing, I learned that my fellow copywriter, Don, likes to collect radio-controlled cars and race them in the driveway on weekends. Oh, and for lunch, Ritchie took me to a new deli in town that sells "overstuffed" sandwiches.

She handed me a piece of paper. "Someone called from *Rolling Stone,*" she said casually. "Here's the number." Then she went into the hallway. "Jay!" she hollered upstairs. "Jay! She's calling!" He crept downstairs and sat next to her at the table.

I dialed with shaking hands, and listened in disbelief as Jann Wenner's assistant, Mary, told me that they would like to offer me the job, and could I start next Monday?

"What?" my mother mouthed. "What?" They stared at me, brows identically furrowed.

I hung up the phone and faced them. "I did it," I said wonderingly. "I got the job."

How to Approach an R & B Artist When You're the Whitest Person in the Western World

Too often, I have seen my pasty brothers and sisters flame out when they try too hard to be d-o-w-n. If you are Wonder-bread white, as I am, and meet a hip-hop artist, do not give a complicated handshake with many different permutations if this is not your practice in ordinary life. If they crack a joke, do not clap loudly, bray with laughter, and holler "That's what I'm *talking* 'bout!" as I watched a white MTV production assistant do to the bemusement of Busta Rhymes. Grover from *Sesame Street* put it best: Be yourself, and if that self makes George Plimpton look like a chocolate funkateer, so be it. I'd rather find my way in with furious research and try to impress artists with my knowledge of the name of their childhood friends or the résumés of their sound engineers.

I once watched an obsequious blond reporter ask the three women of Destiny's Child if they met "in the 'hood." The trio, normally personable, stared at her frostily: Beyoncé Knowles and Kelly Rowland were raised in a comfortable Houston suburb; Michelle Williams grew up in similar circumstances in Rockford, Illinois. No, they said, we did not meet in the 'hood.

When I sat down with them, I did not make that mistake. That said, I wasn't entirely myself, either. Originally, our interview was to take place in Los Angeles, but just in case, I asked the publicist if the group was doing appearances anywhere else. "Well, they're going to Omaha to perform at some high school that won a contest by raising money for charity," she said. Perfect. Off I went to the Millard North High School, where a few thousand white kids boinged off of one another in a frenzy as they waited for the girls to stride out onto the tiny stage. "You need to calm down and be quiet!" hollered the school principal in vain. "No one should be on anyone's shoulders! Feet on ground!" Love that. I wrote that down.

Then they appeared, golden Glamazons resplendent in hot pants the size of a dryer sheet and gold stiletto boots. The kids in the front row, clearly on funkiness overload, had the walleyed look of the Today's Catch section of the supermarket. The trio smoothly ran through a forty-five-minute medley of their hits, and then quickly retreated to their gargantuan tour bus.

My palms were flowing as I timidly approached the bus driver. They seemed so coolly untouchable. "They're in the back," he said. I followed the sound of giggling.

It was bizarre. The gold lamé outfits were dismantled, the makeup was hastily wiped off, and three girls who were barely out of their teens were lounging in jeans and chomping bags of Cool Ranch Doritos and Cheetos with such enthusiasm that the air around them twinkled with orange dust. The disparity between the sophisticated ladies onstage and these clean-scrubbed girls was surreal. For the rest of the day, we had a g-rated slumber party, as they goaded each other into laughing fits. I helped myself to some Cheetos as we compared pedicures and talked about dating. (At the time, they were single, so they earnestly discussed the self-help books they were reading in order to meet the right man, such as *Knight in Shining Armor: Discovering Your Lifelong Love.*) Even though they were blindingly famous, it was all reassuringly familiar territory. A gathering of girls: That, I can do.

We moved on to the topic of cellulite, and then zits. Beyoncé mentioned that she had recently counted the blemishes on her face, and got up to thirty-five. No matter what the topic, they frequently invoked the Lord, holding up a testifying hand when they did so.

Of course, I did, too.

"God has a plan," said Beyoncé. "And God is in control of everything."

"Yes, He does," said I. "Yes, He is." At that particular point, the Creator had every right to strike me down right on that tour bus, because I had not been to church in years. That didn't stop me from chiming in, of course. I was able to remember Bible passages because my folks used to frog-march us kids to church on Sunday, and for years, I sang hymns in Bible day camp, so as the day wore on, I threw in any allusion to the Lord that I could.

At one point, Kelly said that as long as they didn't take their eyes off of God, they would be fine. I nodded in solemn agreement. "Amen," I said. Can I get a witness! I loathed myself. Why did I have to go that extra unctuous mile?

"He will make straight and true your paths," I added.

6.

The second I stepped through the doors of *Rolling Stone* as a real employee, I wanted to shake off my old personality like the rigid husk of a cicada. But how could I cultivate a new, hip persona when I lived with my parents in a New Jersey suburb and wore black leggings as pants?

"You should pack a lunch," suggested my mother on the first few gut-churning mornings before work. "It would save a little money."

"Which you'll need to do," said my father, "because your mother and I have decided that you're going to have to pay rent around here. Fifty bucks a month, and you can pay for dinner from time to time. Nothing wrong with picking up a pizza." My folks could pinch a penny until it bruised. These are people who would water down shampoo so intently that as greasy teens, we pretty much manufactured our own styling wax on our heads.

I was too nervous to fight them. "I'm not packing a lunch," I said. "I want to order it, like everyone else does." This was alarmingly similar to a long-ago struggle: my wish for a *Planet of the Apes* lunch box versus my mom's insistence on a bag lunch. Maybe it was time to start thinking of an apartment, although on my new wages of eighteen thousand a year, it might be difficult.

During the first week of the job, I never questioned why I was hired, just in case it might have been some sort of clerical error. Years later, I quizzed

Bob. As it turns out, we had a lot in common. For all of his urbane veneer, Bob was a scrappy Irish guy from Long Island, the son of an FDNY fireman, who went to SUNY and spent his first three post-college years toiling at a trade magazine for industrial chemicals. But at the time, I gathered that the magazine liked their staffers laid-back and personable, yet highly motivated, which, Lord knows, I was. As an editorial assistant, I eagerly filed documents, transcribed editors' interviews, and—best of all—answered the phone.

In that pre-Internet era, *Rolling Stone* was one of the few sources for music information, so random callers would constantly phone up the magazine with trivia questions. Typically, a group of bankers from Chicago would have a boozy lunch, get in a heated argument about the name of Neil Young's first band (that would be the Squires), and then call us to break it up.

Excitingly, we were often assailed to solve bets, some for serious money. Was the Beatles' "Martha, My Dear" really about Paul McCartney's dog? Well, Martha was the name of his Old English sheepdog, but the song was not about her. What was Iggy Pop's real name? James Osterberg. Was "You're So Vain" about Warren Beatty or Mick Jagger? Carly Simon has never confirmed, but common wisdom is that it's Beatty, while Jagger sang background on the song. Did a Led Zeppelin groupie once have sex with a fish? Affirmative. What were the first few lines of Elton John's all-but-indecipherable "Bennie and the Jets"? "Hey, kids, shake it loose together, the spotlight's hitting something that's been known to change the weather."

Every morning I loved to watch Jann Wenner sweep into the office as if he were coming down the walkway from *Air Force One,* unleashing a torrent of good mornings and brisk nods and pointed fingers ("I'll see that copy before noon, right? Good"). The moment he arrived, his charisma and preternatural energy instantly changed the chemistry of the place and it began to hum and whir. He would usually nod in my direction, but the squadron of young editorial assistants told me that it took a year or so for him to learn your name.

The assistants were pleasant to me, but I knew it would take a while to penetrate the sanctum, so in the meantime I soaked up the surroundings, desperate for clues on how to behave. For a week, my lunch consisted of granola bars from a vending machine because I was still too cowed to order in. Every time a gaggle of employees would converge to talk, I would surreptitiously listen, noting the bright, quick, banter-y way they spoke, as if they were in a sitcom. Around the second week, I felt comfortable enough to indulge in the time-honored employee ritual of buying a backpack from the promotions department emblazoned with the *Rolling Stone* logo. Seemingly every staffer wore some sort of item with the logo on it, so that at quitting time we all looked like runners in the same marathon as we poured out of the building with our *Rolling Stone* bags and caps and T-shirts and jackets. I gloried in the curious, admiring stares that my backpack elicited when I rode the subway.

The music scene in 1989 was dominated by smooth hybrids of pop, R & B, and hip-hop such as Soul II Soul, Neneh Cherry, and Janet Jackson. Pre-Nirvana alternative was thriving (Bob Mould's *Workbook* and the Pixies' *Doolittle* were played incessantly in the office) and the gangsta rap of N.W.A. was stomping on its gentler precursors. Hair metal was waning but still hanging in, sucking the fumes of Guns N' Roses, and cheese pop like Milli Vanilli and New Kids on the Block still had a stranglehold on the charts. The number one song of that year, lest we forget, was "Look Away" by Chicago. A sampling of *Rolling Stone*'s 1989 covers: Madonna (twice), Roland Gift from Fine Young Cannibals, Uma Thurman for the Hot Issue, Axl Rose, Jon Bon Jovi, the cast of *Ghostbusters,* Jay Leno and Arsenio Hall, and R.E.M.

During my third week at the office, I scurried out on my lunch break to buy albums from the Replacements, a pair of Doc Martens, and once, some exotic halvah for a snack. I also added a thorough read of the *New York Times* to my morning ritual after an embarrassing incident during week two. A group of staffers had gathered near my desk, and as they opened their stacks of mail, which bulged with new music, they began discussing Reagan's Iran-contra mess and how it would affect his legacy.

"The Cold War caused more bloodshed than the First or Second World War," said a garrulous research guy. "But even so, remember that Reagan left his presidency with the highest approval rating since FDR. I think schoolkids will ultimately know him as the president who ended Communism. Although he was hardly a genius when it came to foreign policy. He didn't know what to do when it came to Cuba, or China, for that matter—"

"He barely touched the situation with Palestine and Israel—"

I watched them eagerly, absorbing every word. I always wanted to have spirited, informed arguments around the dinner table about politics and ethics and religion like the characters in Woody Allen movies. In college, I imagined that my friends and I would have long, passionate discourses about life and relationships, like people did in Eric Rohmer movies. As I watched the group with wide, unblinking eyes, I thought about last night's discussion in the kitchen with my parents.

It had begun, as it always did, with my father pouring himself a Dewar's. Every night at six he had his Dewar's and a small bowl of something salty like microwave popcorn. He never deviated from his routine, even when I once took my parents to Europe. When I joined them on the first night in their hotel room, I watched as my father unpacked a Tupperware container of Dewar's.

"Dad, you can probably get scotch here," I said. "It's Vienna, not Papua New Guinea."

"Even if you could get it, it's probably more expensive," he said, struggling to remove what looked to be a shiny brown throw pillow that he had shoved in his luggage. It was a large Ziploc bag of Chex Mix.

My father sat heavily down at the kitchen table and got himself situated. "You need to set up a 401(k) as soon as possible," he said.

"Jesus, Jay," said my mother. "She just got the job." Still in her work clothes, my mother kicked off her shoes and reached for the scotch. It was my night to cook dinner, so I was bent over the stove making spaghetti, my old standby. Boil spaghetti, add jarred sauce, cut iceberg lettuce into wedges and slop on dressing, serve.

"She can't rely on Social Security," boomed my father. I could see he was gearing up for one of his favorite lectures: If You Think That Social Security Is Going to Be Around When You Retire, You'd Better Think Again. "By the time she's a senior, she'll probably have nothing," he said. "Payroll taxes would need to double to cover the projected costs of Social Security and Medicare."

Most people harbor one overriding fear, one that both haunts them and drives them forward. Mine is that I'll be old and penniless, which has directly stemmed from the post-apocalyptic tone of my father's seminars.

"How much do you have in savings?" he demanded. I told him the amount. He stared at me for effect, then sorrowfully pulled his bag of popcorn out of the microwave and emptied it into a bowl. "That'll last you a year," he said over his shoulder. "One year."

Our ancient cat, twenty-one and practically freeze-dried, wobbled into the kitchen, sat down, and began to convulsively gag, doing its elderly best to bring up a hairball. "Not on the goddamn rug," my mother said, sighing.

My recollection was broken by the voice of the researcher. "I notice you've been listening to our little debate," he said. "What do you think?" To my horror, the whole cluster of *Rolling Stone* staffers was looking at me with bright, curious eyes. No. Oh, no. This was my nightmare. Why didn't I read my folks' *Newark Star-Ledger* more carefully?

"Me?" I squeaked. I was incapable of this kind of discourse. Usually, I ask a question that incorporates what a person has just said, e.g., "Well, what could Reagan have done about Cuba, in retrospect?" If you pose the question in a strident sort of way, with a slight frown of concentration, you're involved and engaged in the lively discussion without actually saying anything. But in my fright, I had forgotten what was just said.

They all stared. "Don't get me started about Iran-contra," I said heatedly, shaking my head. What I was trying to convey was, *I am so passionately against and so rabidly informed of Reagan's misdeeds that if you get me going, God only knows what I'll do. So do yourselves a favor, and let's not unleash the beast.* They returned to their discussion, and that evening I subscribed to the *New York Times* and the *Wall Street Journal.* In the meantime, until I was sufficiently caught up on world events, I devised a strategy for any sort of staff discussion that was over my head: I became the moderator. If you're the group's John McLaughlin, you can fake being informed while still being involved by deploying a few pointed but vague questions. If a person is holding forth and another is twitching to interrupt, jump in and ask her why she disagrees. Ask follow-up questions. Nod vigorously while saying things like "In what sense?" or "How, specifically?" That way, you smoothly take control of the conversation without actually contributing anything even remotely worthwhile or informative. Before anyone can ask your opinion, remember a phone call you have to return and busily excuse yourself.

While I tried to navigate the social terrain during those first few heady weeks in the office, I had my first celebrity sighting. Jann was friendly with plenty of famous people, so you never knew when they would amble through the hallway. One favorite among female staffers was JFK Jr. He was always very discreet, so a network of female assistants devised a system when he would arrive. "John-John," one of them would murmur into the phone, alerting a coworker further down the hall. "John-John," whispered the next, and so on.

One afternoon as I was opening an editor's mail, I looked up to behold the grooviest trio I had ever seen, undulating toward me in an incense-scented cloud of peace and love: Lisa Bonet and her husband, Lenny Kravitz, who was holding their baby, Zoe, in some sort of batik-printed sling. Lenny and Lisa were easily the best-looking couple of the late eighties: Lisa, at the peak of her sloe-eyed sexiness with glossy hair cascading down her back, was wearing a long, fringed purple skirt, Indian sandals, and a shiny blue wrap top. She was laden with what seemed like the entire contents of one of those peddler's tables that are set up near NYU: strands of clinking necklaces, long beaded earrings, various scarves, and kente cloth bags. Lenny, meanwhile, had dreadlocks down to his shoulders, rose-tinted sunglasses covering his hiply expressionless face, chunky silver wrist cuffs, and striped bell-bottoms that very clearly and unapologetically let you know that he was hanging to the left.

Lenny had recently shed his goofy former moniker, Romeo Blue, and had released his debut album, *Let Love Rule,* under his own name, and Lisa, who had yet to use her goofy new moniker, Loloki Moon, had made the jump from *Cosby* and *A Different World* into film. High hopes were pinned on Lisa after her turn as Epiphany Proudfoot in *Angel Heart.* I gawped at them as they glided toward me, various accessories and jewelry jingling away, and I froze completely when I realized that Lisa was actually going to speak to me.

"Heyyy," she said in her soft, husky voice. "Do you think I could use your phone?"

"Of course," I gibbered, grinning so vigorously that my back molars showed and practically shoving the receiver at her. Her skin glowed as if from the inside. She dialed a number and then took the baby from Lenny. I wanted to give her some privacy but couldn't leave my post because one of the editors was expecting a call from David Bowie, and I had to man another phone line. So there I sat, as she whipped a breast out of her shirt and began to feed little Zoe. Her breast, a few feet away, looked oddly familiar and I realized it was because I had seen it in *Angel Heart* when her character has sex with Mickey Rourke (during that long-ago time, schtupping Mickey Rourke onscreen was a career-booster).

Lisa noticed me staring at her, but she met my gaze serenely, with a look that said, *Go ahead and stare, if that's what gets you off. Everything's gonna be all right. One world.*

She was talking with what seemed to be an assistant about procuring organic food. As she chatted, she gestured with her arm and released a little cloud of incense scent, which I surreptitiously sniffed. Patchouli? Myrrh? I took a deeper sniff. Sandalwood? Chinese Rain?

She hung up the phone. "Thanks," she said with a half-smile. Then she turned to Lenny, who was lingering nearby, and the two of them glided out the door, presumably to go have sex somewhere.

I barely had time to recover from grooviness overload when I had my second encounter with a boldface name. I was walking purposefully past

Jann's office in my new Doc Martens—worn, daringly, with a dress—when I heard a commotion. "I know the answer! I know it!" some guy was hollering in an oddly familiar voice. All of a sudden, Jann popped his head out of the office and looked around. His eyes fastened on me. "Come in here a minute," he said. Then his head disappeared.

Weak-kneed, I walked in and almost collided with the barrel chest of Sylvester Stallone, exalted hero of all of my Jersey comrades, who loyally forgave the release of that year's *Tango & Cash.* At that point, I hadn't realized that most celebrities are smaller than expected, so I was amazed to see he was around five foot nine.

"Your boss and I have a bet going," he boomed at me.

"Yes," said Jann. "We want to know who it was that said, 'Power corrupts; absolute power corrupts absolutely.' "

I looked from one to the other and I cursed, for the umpteenth time, my state school education. How I wished I could just rattle off the answer.

Do something. "I'll find out!" I piped. I scurried out the door and zoomed down the hall toward the research department. "*Bartlett's Quotations*!" I barked to a guy who was quietly reading. Without looking up, he handed the book to me. I paged through it with shaking hands and found the quotation. Lord Acton. It was Lord Acton! Of course!

I raced back to Jann's office and burst in. "It was Lord Acton," I announced in a bemused way, as if to confirm what I knew all along.

"I was right!" said Jann.

"What?" blustered Stallone. "Ah, jeez." Keyed up, he reached over and gave me a playful punch in the arm. I reeled backward and almost fell over, my arms pinwheeling. They both laughed.

"Sorry, kid," he said.

That was my last celebrity encounter for at least a year, until I was given my first assignment, a ten-minute phone interview with Mary Tyler Moore for a special television issue that the editors were putting together. She and I were to discuss *The Mary Tyler Moore Show,* and even though I had seen every episode, I amassed research for days and wrote out pages of questions.

When she called, right on time, I said what I had rehearsed, which was, "Hello, Ms. Moore. It's a pleasure to talk to you."

"Ohh," she said in that lilting, friendly voice of hers. "Just call me Mary."

Phone interviews are odd exercises because you must conjure up an intimacy in short order and with no visual cues. Are they smiling? Frowning? Although sometimes, under the cover of anonymity, both parties can slip into a cozy, confiding tone.

Not in this case. "I saw you in FAO Schwartz a few weeks ago," I said, deviating from my script. Ugh, what was she supposed to say to that?

"Oh, ah, yes," she said helpfully. "I was buying something for a friend's son."

I was so nervous that I wasn't listening. "You were signing some autographs," I said. I was doing the classic, useless, dead-end move of flummoxed fans: pointing out something that said celebrity did. ("Hey, I saw you on the *Today* show.")

"Yes," she said.

Silence.

"You should have won the Oscar for *Ordinary People*," I went on desperately.

"Thank you," she said.

Silence.

I scrambled onto stable ground with some questions, and as she gamely answered I constantly interrupted her so that our conversation was in mosaic form.

"It's interesting that—"

"What?"

"I was just saying that it's—"

"—it's interesting, then? What's interesting?"

"That—"

"Right."

After ten minutes, it was all over. I signed off by telling her it was my first interview. (Another mistake: Hey, seven-time Emmy winner! They sent an amateur to talk to you!) She faked surprise.

"Well, you did wonderfully," she said graciously. I will always be grateful that she was my first interview.

My first prolonged face-to-face encounter was about as far from Mary Tyler Moore as you could get.

"I have a 'q and a' for you," said Karen, one of my editors, stopping by my desk. "He's sort of difficult, but I think you can do it. You'll have to go to his hotel."

I didn't care who it was. "I'll do it," I volunteered, leaping up out of my chair. "I'm in." For the love of sweet Jesus, sit down, I told myself. No need to stand.

"Good," she said. "Here's the number of his publicist. Set it up as soon as you can." She rushed off, distracted.

I looked down at the paper. Oh, God, no. Johnny Rotten. The moody founding father of punk was prone to outbursts, or, worse, would occasionally refuse to talk at all. I loved the Sex Pistols so much that I decided to forge ahead. He seemed to have a sense of humor, anyway. When Malcolm McLaren discovered him, he was wearing an I Hate Pink Floyd T-shirt, and how could you go wrong with someone who once called sex "two minutes of squelching noises"? Still, the memory of his notoriously combative exchange with a hopelessly uptight Tom Snyder on his late-night TV show lingered unpleasantly in my mind.

On the day of the interview, my knees were literally shaking. I timidly made my way to his New York hotel room and knocked on the door, which was ajar. Nothing. After a few more knocks, I pushed it open.

He sat sullenly at the window with his back to me. He had on some sort of plaid suit and his hair stood up in angry yellow tufts. Then he slowly turned his head.

"I hate your mag," he said sourly. Then he turned back around.

"Now, Mr. Rotten—," I began.

"—Horrible," he said in a singsong voice, like a child. "Horrible, horrible, horrible." He stared out the window.

I stood, immobile. What to do? Clearly, he wouldn't care that I was a "big fan." Did he want me to leave? I reasoned that I should act as he was acting.

"Well, you obviously want publicity or I wouldn't be here," I snapped. One of my knees was jumping crazily. "So let's just get this over with."

He stayed motionless. I looked around his hotel room in desperation and spotted a room-service tray that held a half-eaten bowl of cereal and a large container of milk, which I recklessly picked up.

"Wow. Milk," I said. "Pretty punk rock."

To my profound relief, he turned around with a smirk.

None for Me, Thanks: Gracefully Refusing Your Host's Kind Offer of Heroin

Scott Weiland of Stone Temple Pilots was moving restlessly around his cavernous hotel room in Beverly Hills. He sat down. He stood up and paced, unsmiling, then sat down again. Recently sprung from rehab for heroin addiction, he looked far from healthy but still decadently glam in his tight jeans and black T-shirt, his nails polished dark, his near-translucent skin eerily pale.

He was rhapsodizing about heroin, clearly one of his favorite subjects. "I love to do it," he said. "I'd love to do it right now." His long, dry, twiglike fingers plucked insistently at his jeans while he talked. It looked like he could rub them together and a fire would start. "If I had some of it on me right now, I guarantee I could get you to do it in a second," he said, looking me in the eye for the first time.

I told him I was too afraid. He fastened his flat gaze on me with new intensity. "Then I would say, 'Afraid? That's okay. I'll show you how,' " he said. "I've got two clean needles in my pocket. If you don't want to shoot up, you can snort it, or smoke it." He leaned forward. "It's the best high you'll ever get. It's like you're being embraced by God."

My heart was hammering.

"The first time I ever did it, I felt like I found the keys to unlock the doors to all the secrets." He leaned over and inspected a silver room-service tray that sat between us. He selected some roasted vegetables and dropped them into his mouth. "I don't eat very much, you know," he said. His fingers were shiny with oil.

"It goes right to the pit of your stomach," he continued. "Then it crawls up to your lungs, then it rushes to your head. And it's warm, and you just close your eyes and smile. And . . . it's hard talking about it." His eyes searched the room. "I wanna smoke a cigarette or something." He jumped up and darted over to the minibar, grabbed a few vodkas, and pleaded with me not to tell his publicist. He poured them into a cup, glancing furtively at the door and theorizing that if she didn't get too close, she would assume it was water.

He continued to press me to try heroin, telling me in a clear, quiet voice that if I simply said the word, he would order a limo and he could indoctrinate me as the car drove through the Hollywood Hills. This is the paradox of an addict: Ten minutes prior, he had been saying with obvious misery that drugs had stolen his soul and ruined his marriage, that he had turned his brother on to them, transforming him into "one of the undead." Now he was regarding his mesmerized visitor with vampiric greed.

For a moment, I thought about it. If heroin happened to be on your to-do list, there would certainly be worse ways to try it than with a slinky rock star as your limo glided through the gardenia-scented canyons of the Hills. And while he wasn't particularly vigorous, drugs hadn't affected his darkly glamorous looks—aside from his eyes, which were expressionless like two black, shiny buttons.

"Just say the word," he told me quietly.

My mouth was dry. "May I use the facilities for a minute?" I rasped. Normally I would never have asked, but ex-users aren't fazed when you disappear into bathrooms.

I closed the door, ran the water, and called Heather on my cell phone.

"Please advise," I said in a low voice. "I'm in Scott Weiland's bathroom. Listen. He wants me to try heroin."

Silence. "Now?"

"Well, he has to make arrangements."

"Are you insane? Get out of there."

"It's crazy. He's very persuasive. He brought up Keith Richards."

Heather snorted impatiently. "Everybody brings up Keith Richards as the heroin poster boy. You know what? George Burns lived to be one hundred and three, or maybe it was one hundred and two, and he smoked and drank."

"I don't understand. What's your point?"

"My point is that he's a rarity. Most people don't live to be one hundred and two by smoking and drinking. And heroin is only good the first time."

"How do you know?"

"Everyone says that. They say you can never achieve the high that you get the first time. Listen, why are you calling me? Are you really asking my advice? And can he hear you talking in the bathroom? What does he think you're doing?"

I thought about it. "I don't think he really cares," I said.

"My point exactly. Get out."

I emerged from the bathroom determined to take control of the situation. Before he could resume the Wonders of Heroin seminar, I steered him toward the subject of his new album, which sufficiently distracted him. As I quickly wrapped it up and grabbed my tape recorder, he asked me one more time if I was sure I didn't want to try it. For the first time that day, the corners of his mouth twitched into a smile.

"I'm sure," I said.

When I got back to my own, decidedly less elaborate hotel room, I did what I usually do to calm myself down: I called my father. He was, of course, always willing to dispense advice, but his actual words were less important than his overall tone. Even the way he answered the phone was in the classic, genial manner of every small-town hardware store owner or Old Spice–

scented Rotary Club guy at a pancake breakfast: *And what can I do for you today? Uh-huh! Well, I think that can be arranged!*

"Hello?" he said expectantly, cheerfully. His manner is classic Midwestern Hearty (broad, accessible, pleasant comments laced with wry observations about family) as opposed to Southern Hearty (flirty, teasing) or Northeastern Hearty (remote until the booze kicks in). If I answered the phone at all, it was with suspicion and dread. Not my father. No caller I.D. for him. He'd rather be happily surprised. Maybe it's one of the girls! Or Vern Leister, my old buddy from Penney's! Wonder what he's up to! Or maybe the folks at United Way, needing me to help out this Saturday for the charity golf tournament! Well, I think that can be arranged! Wrong number? No problem! Have a nice day, now!

"Hi, Dad," I said, shrugging into a hotel robe and clicking on the TV's movie menu.

"Are you in Los Angeles? How did your interview go?" I could hear Lite-FM music playing in the background. My father lived in a soothing, wall-to-wall-carpeted world where the thermostat was rigged on a timer. Everything was easy. Even their gas fireplace flamed up with the press of a button.

"He was a little unhinged. I'm fine, though. What are you doing?"

"Oh, I'm just putting pictures in a photo album and paying some bills. Your mother's outside gardening. Her roses have beetles and she's on a mission. And then, let's see, later we're going to watch a movie that I rented, something Tim Wiggin recommended."

My blood pressure began to drop. This was just what I needed. "What did you rent?"

"I knew you were going to ask me that. Let's see. Uh . . . it's in the living room somewhere." There was the sound of him rummaging around. "I never know where your mother puts things. Oh, I don't know where it is. I think it's called *Little Foot.*"

Movie titles were not my father's strong suit. "I can't say I've heard of *Little Foot,*" I said carefully. "Do you know who stars in it?"

"Oh. Jeez. The guy who was in that crime show in the seventies. You know."

"What's the plot?" I prompted. "Is it some sort of Native American drama?"

He thought for a moment. "It might be. No, wait, it isn't. I think it's about adoption."

"Well, it's not important. Are you two okay?"

"Oh, sure," he said. "You all right, honey? Are you having some room service? You should relax. Treat yourself."

"I will," I said.

I didn't tell him about my encounter. Why worry him?

7.

The more I was exposed to the gloss of New Yorkers and famous folk, the more boorish Ritchie's behavior seemed. Once I started settling in at *Rolling Stone,* Ritchie had an expiration date on his forehead. The final straw was the company Christmas party, which was held at Nell's, the legendary nightclub that I had hungrily read about in my college dorm room whenever my issue of *Interview* arrived. By 1989, the club was a tad past its prime, but not to me.

I had only been on staff for a couple of weeks and still didn't know anybody that well. "Maybe I should stay home," I fretted to my parents.

"Oh, just go," said my mother. "You can network."

"I'll bring Ritchie," I said. "That will make me feel better." My mother raised her eyebrows at my father.

Ritchie wanted to drive into the city rather than take a more sensible train. He arrived at my house two hours late and wearing his favorite "party" trousers, which were modified parachute pants. "I told you to wear jeans," I said.

"What, this isn't good enough for your New York friends?" Lately, I got a lot of those remarks.

"No, no, it's fine," I said.

I caught my breath when we walked through the door. There were the velvet couches and the crystal chandeliers, just as I had seen in so many pictures! And there was queenly Nell, with her cap of glossy bobbed hair, kissing cheeks and calling "Hello, darling" over the throbbing De La Soul music. Nell, who made everyone, no matter how famous, pay five dollars to get in! I was in Nell's, hangout of Mick Jagger and David Bowie!

"They once turned away Cher at the door," I explained to Ritchie.

"She's a slut," Ritchie pithily observed.

We walked around, taking it all in. "Nell used to dance naked on the tables," I said.

Ritchie chortled. "Huh," he said with a leer. "I wish she'd do that now."

A lush buffet was spread luxuriously around the bar, and the waitstaff, all British, passed around endless trays of champagne. *Rolling Stone* often featured entertainment at their parties—one year, it was said, the actual Rolling Stones played—but not on that particular evening. Still, the holiday bash was appropriately extravagant.

I shouldered my way through the crowd toward one of my fellow assistants, a funny, affable guy named Chris. He was standing with a group of the younger editors.

"Hey," I said happily.

Ritchie came up behind me. "Where the white women at?" he hollered. This, a quote from *Blazing Saddles,* always got the festivities rolling at our Jersey parties. Everyone gave him a pained courtesy smile. I watched as Chris gamely tried to engage Ritchie, who was draining his fourth glass of "free" champagne. Why did I ever think he was a cheekily dangerous bad boy?

Five glasses, six glasses, eight glasses, ten. "Maybe we should go," I said. "And you should stop with the champagne, because you have to drive home."

"I'm a great drunk driver," he said. "You know that. When I've had a few, I put all my concentration into it, so I'm more careful." He made his way over to one of the plush velvet couches, where he did a somersault and tore his parachute pants.

Nell, who had been talking to Yoko Ono, came rushing over. "Hey," she said. "Don't do that."

I broke it off with Ritchie that night as his car idled in my parents' driveway. ("What?" he said, his high spirits deflating abruptly. "What?") I missed him for a few weeks and contemplated calling him whenever I had an awkward moment at work, but once I saw him in a different light, it was over. And my recovery was sped along because at *Rolling Stone,* there was always a party going on somewhere.

Every night after work, I joined the conga line of employees who rolled out of the office for drinks, and then dinner, and then off to the front of the line to see Eleventh Dream Day or Jesus and Mary Chain.

The daytime wasn't off-limits, either. I watched with envy as some of the editors stumbled in, loud and red-faced, after a boozy lunch. After hours, the bathrooms were once a popular place to Hoover up coke (helpful for making deadlines), and I surprised a couple of revelers more than once. One night, a staffer, relieving himself in the john, looked down and noticed a neat mound of white powder on top of the urinal. Ah, he thought. Somebody had a party and hastily ran off. He looked around to make sure he was alone. All clear! He licked his finger, dipped it in the pile, and rubbed it vigorously on his gums. It was Ajax.

I stuck to booze because it was all I could afford, and I drank a vat of it every night before catching the last bus to New Jersey, dozing during the ride home as the alcohol gave me raisin eyes and a cat tongue. During my garish nighttime carnivale at various East Village bars, the memory of Ritchie dropped cleanly away. Not that I had any other prospects. Typically, I would be approached in the following way:

SETTING: King Tut's Wa-Wa Hut, a cramped bar on Avenue A. As I wait for a drink at the bar, I notice a lanky, dark-haired guy leaning woozily against the wall. He had all the totems of late-eighties hipsterhood: jeans with two perfect rips at the knee, the requisite drooping forelock à la Johnny Marr of the Smiths, and one tastefully dangling earring on the "straight" left ear. His eyes were glittering slits, and he appeared to be wavering in and out of consciousness.

HIPSTER, MUTTERING THICKLY: Ni——f'you . . .
ME, LEANING FORWARD: Beg pardon?
HIPSTER, MORE LOUDLY: Can I fuck you?

I stared at him. Well, you had to hand it to the guy. Why not cut right to the chase? Here was an admirably straightforward young man who had no patience for chitchat or the silly facade of buying drinks! I half-contemplated taking him up on his offer.

Even if I could get a date, I would have to break it to them at some point that I still lived at home with Mom and Dad. Although I was reluctant to be pried out of their house, the urge to go to a bar without having to catch the 11:20 bus to Jersey became too strong and I made the bold move of finding a studio apartment in Hoboken, the "mile-square city" made up entirely of kids from New Jersey who recently graduated college and couldn't quite make the move into Manhattan.

My last night at my folks' home was a mournful one. We had our usual Sunday dinner of steak and potatoes, which we traditionally ate in front of the TV so we could watch *60 Minutes.* ("Well," my father would inevitably announce after the exposé du jour, "that certainly makes you think.") Afterward, I snuggled into my bed with its crisp sheets (Sunday was laundry day) and lay staring into the darkness. No more Sunday dinners, I thought sadly.

I eased the transition by arranging to go on my first date that wouldn't end in a frantic scramble to catch the last bus. As a bonus, he actually hailed from New York, not New Jersey. His name was Josh, and he was a friend of a music publicist whom I knew who was intent on setting us up. He grew up on the Upper East Side, bouncing merrily in and out of prep schools.

"You'll love Josh," she said. "He's so fun." Fun. Red flag, ahoy. I smoothly ignored it, as it had been a six-month post-Ritchie dating drought.

"What does he do?" I had learned to ask that question first, like a true New Yorker.

She shrugged. "He's really rich. I think his grandfather invented waterproof fabric. He's trying to start a p.r. company for nightclubs, but most of the time he just sort of bops around."

"Would you sleep with him?" I always asked that. Most of the time the answer was a stammered, "Well, no, he isn't really my type, but he's a supernice guy and all the girls in my office think he's hilarious, and—" At which point I politely declined. I was constantly approached by well-meaning friends who wanted to pair me up with the asexual brother type in their workplace, the one who never had a girlfriend but was *so sweet* and *really very attractive*, a benign, pleasant druid with B cups who told corny jokes.

She laughed knowingly. "Actually, I did sleep with him," she said. That seemed weird to me, but I kept silent. "I'll have him call you."

He picked me up at my Hoboken apartment in a red Porsche. A more sophisticated woman would have rolled her eyes at this flagrant display of Sacagaweas, but for a suburban girl weaned on John Hughes movies and their emphasis on shiny sports sedans equaling the Good Life, this was exciting. Plus, where I came from, a date picked you up in a car. How else were you supposed to get to Fuddruckers?

Josh had curly blond hair, a dash of freckles on his nose, and, from what I could tell by the slant of his eyebrows over his Ray-Bans, a mischievous expression. "Hello," he said familiarly and kissed my cheek. I jumped in the car and he stomped on the gas. "I thought we'd start with a drive through Central Park," he shouted over the Scritti Politti tape he was playing.

We zoomed through the Holland Tunnel as he hollered some questions. "So! You're a rock writer!" he said. "Who have you interviewed lately!"

I launched into the story of my latest encounter: Earlier in the week, I had gone to John Waters's unexpectedly tasteful West Village apartment. He immediately took a Polaroid of me, which is his custom for everyone who passes through his door, and added it to a large photo album of gamely smiling deliverymen holding bags of food, various friends, some celebrities, and a few confused repairmen. He was a dream interview, whether he was talking about his favorite scene in the campy movie *Anaconda,* "where the snake pukes Jon Voight," or mentioning a recent book signing in which a fan pulled a bloody tampon out of herself to sign. Which he did.

"He was saying that he doesn't like all the new drugs, like Ecstasy," I shouted.

Josh smirked. "Why not?"

"He said that he was in England when the whole country was on it and it was scarier than the Summer of Love."

"I happen to love the new drugs," he shouted. "And, for that matter, the old drugs. In fact, I just did some fat rails before I picked you up." I didn't want to ask what "rails" were, but judging from the way he nearly took out a jogger as he careened through the park, I guessed it was coke.

An hour later, I found myself in a nightclub, where Josh batted away my questions with charmingly evasive answers and hounded me to take some Ecstasy that he just happened to have in his pocket. "You should know about the things that you write," he wheedled. He brandished a white pill. "Try it, you'll like it."

Well, why not, I thought. At that point, I still didn't know whether I liked Josh. At least this might make the night more interesting. I wondered why my publicist pal thought we would be a good match. Already, I was learning how to entertain with anecdotes about my job. Hiding behind other people's punch lines was easier than divulging anything substantial about myself, and most people were content not to delve any deeper. Was I fun, too?

I grabbed the pill and swallowed it down. "That's better," he said, grinning. Then he gulped two pills of his own. "They should kick in shortly," he instructed. Then he leaned forward. "So. Rock Chick. Tell me about yourself." He gave my shoulder a little shove. It was charming in a third-grade sort of way. "What's in the drawer of your bedside table?" Ah. I gathered that for the getting-to-know-you portion of our date, he wasn't going to take the conventional where-did-you-go-to-school approach but was trying the less-traveled route in which seemingly insignificant questions produce a truer sense of what you're All About. I knew it well. I just wasn't able to deliver the flirty answer that was required of me. My bedside table contained cuticle cream, which I dutifully applied every night, and a collection of stories by Sarah Orne Jewett. Oh, and a letter from my grandma. "I know you meet some rough types in your job," she wrote on a note card that had a bluebird with a letter in its beak, "but I know you are"—this part was underlined twice—"*my own sweet granddaughter.*"

Which was actually the case. Why was I in this club doing Ecstasy with Josh What's-his-name when my actual interests more accurately mirrored my grandma's? I was All About gardening and baking and films that featured indomitable middle-aged heroines who take tea on rainy afternoons in Cornwall.

Josh was staring at me expectantly. I supposed I should just get it over with and say that my bedside table contained the *Kama Sutra*. He noticed my troubled expression and smoothly switched gears. "What are you interested in? Let's hear it." He was looking at me with radiant intensity. Maybe he wasn't as shallow as I thought.

"Well," I began haltingly. "I guess you could say my interests are a bit esoteric."

He put a reassuring hand on my arm. "That doesn't scare me," he said warmly.

I cleared my throat. "Since you asked," I said, "I guess lately I've been doing a lot of research on the death of Charlotte Brontë. There have been conflicting theories as to how she died." I snuck a look at him. He still seemed alert. Okay, then. "She died in 1855, and her death was listed as something called phthisis, an archaic term for tuberculosis—remember, both of her sisters died of the disease." He nodded. The music in the club began to throb more loudly. Why was I talking about Charlotte Brontë in a sticky-floored club on a Friday night? Normally I edited myself, keeping it quippy, light, and focused on the other person, but I found I couldn't stop. I just felt . . . safe. I would peel back the layers with Josh!

"Others say that she missed her sisters and willed herself to die," I continued breathlessly. "Some Brontë scholars say that she contracted typhoid from her old servant, Tabby, or that she was pregnant, and had a bout of throwing up that was so violent, and relentless, that it killed her in short order." Here was where I really got rolling, because Victorian-era diseases really turned my crank. "You could easily become gravely ill with a lot of vomiting, because you weren't able to keep fluids down, and it's not as if they could administer a drip back in the day. You know, Josh?" He nodded. Did he know? I found I didn't care.

I thought for a minute. "I've been researching this, and it seems to me that it was a combination of these things—maybe she was infected with consumption, complicated by a pregnancy, followed by a gastric infection, which finished her off." I couldn't stop now. "She always had lousy health, and surely she was infected with TB, because she came in such close contact with her sisters. She shared a bed with Anne right up until the end."

"Why, exactly, are you researching this?" he asked.

I looked at him shyly. "Oh, I just went through a Brontë phase and reread all of their novels. I tend to binge." He laughed. He was nodding at me with that same kindly look. He's smiling! He doesn't think I'm a weird sad sack! I had heretofore shared my Charlotte Brontë death conspiracy theories with only my bosom chums. Maybe I could regale him with mesmerizing facts about my other recent obsession, the Spanish influenza epidemic of 1918. Maybe he was unaware that it killed 675,000 Americans, more than the death toll from all of the twentieth-century wars combined? Perhaps it might interest him to know that FDR, Woodrow Wilson, and silent screen legend Mary Pickford survived it? A tiny, still-lucid part of my brain knew that the drugs had surely kicked in, because any sober person would recognize that any mention of Mary Pickford was pretty much a date-killer.

Yet he seemed enthralled by everything I said. Or was that a vacant look? Then I realized he was gazing at me with Concern Face, the very method I use on celebrities. I paste it on when one of them is prattling about their regressive therapy or how yoga centers them or how they are coping with the recent death of their bichon frise. You regard them with kindly, twinkling eyes, while nodding with a benevolent smile. I care. I do. Go on. Please.

Josh was on a double dose of Ecstasy. Of course I was interesting. "Let's get out of here," he said abruptly, grabbing his coat. As we walked a few blocks to his parked Porsche, the Ecstasy pulsed through me in a shimmering wave. Wasn't I supposed to feel happy?

"I think my skirt is falling down," I said, too rapidly. We passed a sidewalk café. Everyone was staring at me. Was my skirt around my ankles? "Josh," I said. "Is everyone staring at me?"

"No," he said soothingly.

"Is my skirt falling down? Josh. I think my skirt is falling down." My words were piling up against each other. He turned around and put his hands on my waist. "Your skirt is on. See it? Your skirt isn't falling down." We walked on. Where was the car? Where was the car? Where was the car? My foot caught on something. I suspected it was my skirt, which was clearly bunched around my ankles. Was that homeless man staring at me? Is that a ringing in my ears? Josh sighed grimly. "Let's go to my place," he said. "You'll feel better." It was a flimsy premise, but I took it. I had to get off the street.

His apartment, a fifth-floor walk-up, was cramped and smelled like cabbage. "I thought you were rich," I said with my newfound drug-induced candor.

He grinned. "Most of my money is held in a trust until I turn thirty," he said. "I guess my family doesn't trust me. But I don't really spend much time at home, anyway." He went into his bare kitchen and opened the fridge. "I like to be out," he said absently, rummaging through the fridge for beers. "Out and about." As I waited, I snuck a peek into his bathroom. An open magazine lay on the floor across from the toilet, conjuring up an unwanted visual of a multitasking Josh. Through the semidarkness, I scanned his small apartment, looking for clues about my elusive date. Where were the books? "Here's the living room," he said, handing me a beer. Then he pointed to the bedroom. "And here—"

Don't say it. Please don't say it.

"—Here is where the magic happens."

Right. I spotted a lumpy figure on a chair in the hallway. I squinted. It was a Spuds McKenzie stuffed animal. I ran over and gave it a few satisfying punches. Evidently the drugs were wearing off, because I was starting to feel reassuringly hostile once again.

"Hey," Josh said sharply. "Hey."

He guided me onto the couch. "Have a seat," he said, patting a cushion. Then he leaned over and inserted his tongue in my mouth. "You have the greatest dimples," Josh murmured. I didn't have dimples. His hand slid from my shoulder down to my stomach. Then he stuck a finger in my belly button. It stayed in there for a few seconds, but when someone's finger is lodged in your navel, lightly rummaging around with no obvious purpose, those seconds stretch rapidly into weeks.

I can't do this. This was the kind of debauched evening that I had eagerly read about in *Less Than Zero.* Why couldn't I just play along? Josh was cute, if opaque (did he keep calling me Rock Chick because he couldn't remember my name?), and despite his disquieting attachment to Spuds, had a kind of insouciant charm. I suddenly felt overwhelmingly sad. I wanted, if I was being honest with myself, to be home in bed. Alone.

I clutched wildly for my purse. I needed oxygen. As I broke for the door, John Waters's words came back to me. "Who," he had said with a shudder, "would want to *love* everybody, on Ecstasy?" Drugs required an abandon that I just didn't have. Either I became green and fetal, twitchy and paranoid, or filled with gloom for the whole human condition. Some Rock Chick.

The Contrived Activity

Have you ever noticed how many profiles start with a starlet nibbling on a salad at a restaurant? That is because 90 percent of celebrity interviews take place in a restaurant in Los Angeles, usually a quick drive from your subject's house or manager's office. Some magazines have a ban on the dreaded restaurant interview because it's so clichéd, but few have the power to actually enforce it, so usually the only "color" you will get is a recitation of your subject's lengthy dietary requirements to the fawning waiter. ("Well, is there oil in the dressing? Are you sure? You know what, can I just have lemon juice on the side? I'm thinking there's probably dairy in the corn chowder, and that's really, like, *not cool.* And I'm allergic to nuts, so do you use any peanut oil? Could you please please please just ask the chef? Theeenk yew.")

If this is the case, use the classic writer's trick of starting the piece with a dramatic event in your subject's life. That way you have a grabber for the first couple of paragraphs, and then you can ease into establishing the scene in the restaurant and how your subject picks at a plate of steamed kale. If the person is in the midst of living down a drunken episode or custody battle or rehab recidivism, by all means commence with that, even if he offers a "no comment."

Like so: "It's hardly a secret that last month, Mr. Star was caught scraping up a half-gram of coke that he had dropped onto the men's bathroom floor of his favorite local strip club, Titz."

"My lawyer told me I can't comment," he says, digging into his pear and Roquefort salad at a Santa Monica restaurant. See how the boring setting was slipped in there? Who would even notice?

If the person is not enmeshed in some scandal and only has a record or a movie to promote, pry an answer out of them that could serve as a lead. Have they ever stolen something? What was the best day of their lives? (This works especially well for a heartwarming, soft-focus profile. "Mr. Star will never forget his tenth birthday, the greatest day of his life. His father was still alive, the girl he had a crush on was coming to his party, and he was years away from his first snort of coke.")

Another question that can occasionally elicit a dramatic, lead-worthy answer is "When was the last time you were completely alone?" This should only be posed to the upper-echelon famous, the ones with a fleet of minders and omnipresent security. Often, they have not been by themselves for months, or even years. "When I go to the toilet" is a common response. This is one of the strangest things, in my mind, about being famous. How can you never have any contemplative time to let your thoughts range over hill and dale? How are you ever able to fully recharge if there is always someone around you? I asked Madonna this very question and her response was "Twenty years ago." The only time in that period that she was well and truly alone was during a vacation in Greece with her family. She paddled out in the ocean on a raft. Then she paddled back. That was it.

If you can manage to steer your celebrity's handler away from a restaurant, you must cook up the Contrived Activity, in which the two of you will go to the dog run, or play miniature golf, or do laundry—anywhere but an "eatery." The thinking is that if a famous person is distracted by an activity, he or she will magically open up and chatter away, free from the tyranny of facing you across a table.

If you get a few moments of "walk-around time" after a restaurant meeting, think fast. I was once grilling Cameron Diaz at an appealingly ratty New York burger joint called the Corner Bistro. Afterward, we were scheduled to walk around the West Village together, which was about as relaxing as it seems. She was friendly and pleasant but not warm, like a fluorescent light. As I matched her long strides, while trying to make it seem natural and easy that I was holding a tape recorder near her mouth, I scanned each block in a mild panic. What could we do? You must always be plotting and planning, because you usually get an hour or two to spin into five thousand words. For activities, you need a Plan B, C, and D.

We ducked into a bar for a quick drink, but a bar is almost as glaring a cliché as a restaurant. As we walked onward, I tried to tamp down my anxiety because I could tell she was getting a little impatient. Aha! I spotted a fortune-teller's window. Would she like to get her palm read? "Sure," she said. Fabulous! Oops, not so fabulous: She wouldn't permit me to come in with her, so I waited outside. When she came out, she wouldn't tell me what the soothsayer had said. Another bust.

Just when despair set in, I heard some loud music throbbing in the distance. How could I have forgotten that it was Gay Pride week? It was almost time for the parade! I steered her quickly toward Christopher Street, and I will always be grateful to the float of gyrating Asian men in scanty geisha costumes for providing me with a much-needed scene.

When I met with the Olsen twins in Los Angeles ("Please call them Mary-Kate and Ashley Olsen," entreated their rep), our first endeavor was useless: the photo shoot. In most articles, you can't really place the action in a photo shoot because then it's painfully obvious that you have been wedged in there to save time. Also, the majority of photo shoots are not that exciting. Most of the ones I've attended take place in the morning, and they run most of the day, while tons of people stand around with their arms folded and watch the photographer and his assistants, or duck into the hallway to make cell phone calls.

When I arrived at the warehouselike studio, the photographer was setting up and the twins had just arrived in their separate Range Rovers: Mary-Kate first, the "bohemian" one with an armful of rubber bracelets and hoop earrings with birds in them. "Hieeeee," she said in a reedy voice. I commented on her earrings. "These are my doves, my love doves," she said. "I got them on the street in New York." Even according to Hollywood standards, she was absurdly tiny—five feet tall and emaciated, with her kid-sized sweatpants hanging off of her bony butt. Then Ashley walked in, the more polished, reserved twin, also wearing sweatpants and flip-flops. They were both beautiful but completely sexless.

Simultaneously, they spotted a Ping-Pong table in the lobby of the warehouse. Without talking, they made their way over to it, picked up paddles, and silently began to play, their faces blank. It was hypnotic to watch these mirror images wordlessly lobbing the ball back and forth. *Clip-clop. Clip-clop. Clip-clop.* Maybe that could be a kind of scene, I calculated. As they were called over to try on clothes for the shoot, I watched them smoothly snap into gear, as they have done since birth. In a stroke of genius, the photographer wanted to dress them as Brooke Shields in the famous eighties Calvin Klein ad. As the camera clicked away and the wind machine amped up, blowing their hair back, they pressed up against each other, and stared at the camera, unblinking. During a break, they teetered in high heels over to the craft services table. Ignoring the sandwiches that were piled high on a platter, they clutched some nuts and nibbled on them, looking like the world's most glamorous squirrels. The scene wasn't much to go on, but I was still to meet them the next day at the requisite crunchy L.A. restaurant.

As we stood in line, the two of them scanned the menu. "What's gaz . . . gaz . . . ," said one, her forehead wrinkling in confusion.

"Gazpacho?" I said kindly. "It's a spicy tomato soup, served cold." Then I added, absurdly, "You wouldn't like it." How the hell would I know if she'd like it or not? But she nodded in agreement.

Afterward, our contrived activity was a shopping trip to Lily Et Cie, a vintage emporium favored by the red carpet set. I drove with Ashley, who could barely see over the dash of her enormous Range Rover. Ashley kept phoning her sister in her corresponding Range Rover, because the two of them weren't sure of how to get to the store. "Okay," Ashley would always sign off when her sister called, "love yoooou." Good stuff, thought I. She rounded a corner a little too sharply. How hideously fitting, how beyond ironic, if I were to get in a fatal car accident alongside one of the Olsen twins. People would trample on me in their rush to help her, and then I would be consigned for all eternity as the "unidentified woman" who died next to Ashley Olsen.

After finding the place, they flanked me on the sidewalk as we headed for the entrance. Between them, I felt like a lumbering wildebeest, galumphing down the street. They were so delicate, with their huge sea green eyes and their cornsilk hair, little fairies lifted from an art nouveau print. As we entered the store to great fanfare, a very large man in a plaid shirt suddenly materialized. He briefly conferred with one of the girls and then took her keys.

"Who was that?" I asked their publicist, who said it was one of the girls' bodyguards.

Huh? "I haven't seen them once," I said.

He nodded. "Exactly. They're supposed to be invisible. The girls want to keep things as normal as possible."

I never saw them the entire time. Presumably they leaped into a hedge or something if one of us looked around.

The point is: A contrived activity can yield hidden gems. I certainly didn't use anything from the session at the store, where I stood around while they tried on dresses, but I did rejoice at the stealthy bodyguards.

8.

What most people find festive—a weekend at a beach shack with friends, a boat trip down a river, a crackling bonfire on a summer night—I see as a bleak nightmare to be grimly endured. I would sooner put lit cigarettes in my eyes than share a vacation house with a crowd. Inevitably there is one bathroom for ten people, so there is a constant line, and when it's time to do your business, someone outside of the rickety door decides at that moment to take the CD out of the player as you furiously pull up your pants in the silence. Later, you are thwarted again as you realize that if you can clearly hear your friend's newspaper rustling as he reads the *Times* out loud for everyone's amusement, then they can all hear you. The days crawl by as you swell like a tick. No, thank you.

I do not want to stand in the kitchen with the car keys, seething, while one person makes a grocery list and another hunts for cash and a third announces to the housemates playing touch football that all fourteen of us are going to the grocery store in one car for a shopping expedition that should take ten minutes but will stretch for three hours, do you want to come along?

Every eternal day revolves around the meal. If you're at the beach, there's always someone who feels that it's their duty to boil lobsters, a joyless process of liberating the creatures from their muddy prison at the fish market, praying for the water to boil so they'll stop struggling, mustering your appetite as you wrestle the meat out of the shell, and then cleaning up the carcasses, the stench of which hangs over the kitchen for the remainder of the week.

If you're in the woods, you try to devise a menu from the macaroni and cheese mix and Vienna sausages offered by the bait and tackle shop that also sells toiletries and food, or, with noisy fanfare, you open the spider-corpse-encrusted grill out back to barbecue some dubious meat, which will be cold and raw in the middle and burned on the outside. On another night, you will make spaghetti, which the cook keeps tasting with the same spoon and putting back in the sauce, and you can count on someone throwing the cooked pasta against a cabinet door to see if it sticks, done to much hooting and clapping. When it comes to meals, everyone pitches in, so that your food is lovingly touched by fourteen sets of grimy hands, and since everyone is usually drunk by cleanup time, there will always be at least one chunk of beige food stuck in your fork tines when you eat something the next day.

The mantra of the gathering is always "Do your own thing," but of course you can never really do your own thing without acute self-consciousness. If you bring up a book that you're dying to finish, someone will plop down next to you and ask about what you're reading, or a group will gather around you and talk loudly so that you read the same paragraph three times. Somebody always brings a dog, usually a black Lab, and no matter how carefully you edit the guest list, there's inevitably one really annoying person in attendance, either some girl who gets too drunk and cries, or a meathead who likes to repeatedly remind her about it the next day when he's not checking all the various sports scores on TV as the birds chirp merrily outside. You buy flowers at a roadside stand to decorate the house, and in the tumult, nobody puts them in a vase. Days later they've turned to mulch on the counter where you left them, buried under a mound of moldy kitchen rags.

Silence is not going to happen, because silence doesn't mean Good Times, so there's constant chitchat, and one guy who takes it upon himself to play deejay. After lunch, time halts completely and gets stuck at four thirty for what seems like days, so the whole cabal bumbles around until someone cracks a beer and everyone else, relieved, follows suit. Then it's time to go to the grocery store.

After dinner, you can't go to bed early because everyone feels compelled to do the late-night *Big Chill* thing, and besides, there's an uncomfortable undercurrent because one couple claimed the "good" bedroom, despite having just joined the group this year. Then it's activity time. No, thanks, I don't play cards at home, so I sure as hell don't want to do it here. Or Boggle. Or charades. But you finally give in, and you drink more than you want to, and Boggle starts to seem sort of fun, and you think, *Hey, this isn't so bad.*

But then the next morning, after a restless, sweaty sleep on yellowed sheets and a musty dog-hair-covered afghan that the original house owner's aunt knitted during the Eisenhower administration, you jolt awake at dawn to the sound of the stereo blasting courtesy of the one early-riser guy who's annoyed that no one else is up after he has already run five miles on the beach. Fuzzy headed, you make your way downstairs, where there is always a person eating cereal and making chipper small talk before you've had your coffee in a seventies earth-toned mug that's cracked and glued back together and has an ancient lipstick mark that has never been washed away. You grab the carton of warm orange juice that a housemate has left out on the counter overnight and pour it into a glass that foams up from the dish soap that somebody forgot to rinse during the drunken group cleanup.

Then, all you want to do is bike into town to that quaint little scone shop that you spotted during the drive in, the one that looks like an English cottage with morning glories covering the sun-dappled front patio, and buy yourself a scone, a cappuccino, and a newspaper and quietly read, but that is not what this weekend is about. Because even though the unofficial motto is "Do your own thing," if you actually do break away, there are raised eyebrows and hurt feelings, or, worse, as you make your escape and pedal desperately to the scone shop, you discover that you're playing Follow the Leader to fourteen bikes. Then your boisterous, hungover mob noisily overwhelms the tiny scone shop. All the gentle regulars flee as the girl who drunkenly cried the night before complains that the store doesn't offer soy milk and the whole posse rearranges all the tables with loud scraping noises, so that everyone can sit together. God forbid you have two newspapers.

When you can't put off taking a shower any longer, you wonder why you didn't bring your flip-flops as you behold a rainbow assortment of pubes on the floor of the mildew-scented stall. After you're done lathering up in a trickle of cold, rusty water with Prell—always Prell shampoo, bought from the local tackle shop that sells toiletries and food—you reach for your one towel that you had carefully placed on the third hook, only to find it in a wet, fetid pile next to the john after it has clearly been used to swab your friends' nooks and crannies.

Your mind races. Who used the shower before you? Was it one of the clean ones? Was it one of the guys in the nice gay couple or was it the husky one who came out of the bathroom after breakfast cheerfully announcing that he needed a plunger? Who is having actual fun here except the meathead guy, and the couple who don't have a good relationship and are just relieved to be around others? As you prepare to go on a communal trip to the ancient movie-rental place that has *Jaws* in the New Releases section, and the long debate commences as you all try to find the one movie that hasn't been seen by all fourteen of you, you vow to yourself, *Never again. Never, ever, ever.*

So why I thought a hayride would be any different, I don't know.

A new friend from *Rolling Stone* had invited me for an upstate idyll, and in my eagerness to be included, I had ignored the red flags. Every year the family held a hootenanny at a farmhouse estate. I loved farms. Maybe it would be lambing season! What could be the problem?

As it happened, it was the roster of events, which would seem like great fun to anyone else but was, to me, the lowest depths of misery: a tennis tournament, a square dance, communal sleeping quarters, and a hayride. And so I found myself sitting on a scratchy bale of hay, bouncing over pristine farmland in a truck driven by a stalwart farmer type. The celebrants around me, most of whom I didn't know that well, were throwing hay at one another, tossing back drinks, and occasionally bursting into song, a living, squirming, shouting Ralph Lauren photo spread. I, meanwhile, was calculating how long this trip could reasonably last. Two hours? An hour? The tractor had to run out of gas, eventually. Why wasn't anyone looking at their watches?

I surveyed the ring of faces. All were merry, pink-cheeked, chatting animatedly.

Except one. She had wedged herself into a corner, and her ghastly isn't-this-fun smile matched my own. She had the same careless blond good looks of the privileged people around her, but somehow I sensed that she was approachable. I made my way over to where she sat, fighting to keep my balance as the truck heaved over another rock.

I assumed a pleasant expression. "I haven't been on a hayride in years," I said. I tried to be upbeat, but that was the best I could do.

"I never have," she said. "Jews don't do hayrides."

I scrambled to sit closer to her. "This may be the most awful day of my life," I said in a low voice.

"I was once in a car that caught fire," she said. "This is worse."

Julie was a high school friend of the host. She had been visiting her folks nearby and decided to stop in. "No matter how much I drink," she said, "I'll never attain the level of drunkenness to appreciate this." She told me that she lived on the Upper West Side and had gone to NYU film school. She was single, dating here and there, and she wrote scripts for the *National Geographic* Channel.

"I'm a reporter for *Rolling Stone,*" I said. This was not strictly true. I had only recently become an assistant editor.

"I know a guy who works there, named Peter Sloane," she said. "I took riding lessons with him." Ugh. Smarmy Peter Sloane, Mr. Ski Tan. Mr. "Can I get you ladies a drink?"

"Oh," I said carefully. "I know him." Then I cracked. There was something about her that made me want to drop the facade. "He's . . . he's . . . what's the word?"

She laughed. "How about 'horrible'? There are three people I hate in the world. Frank Stevens, Providence Insana, and Peter Sloane." I didn't have time to ask about Frank or Providence as the conversation bounded along.

She leaned forward. "Listen, I don't work at *National Geographic.* I just sent them in a test script. I don't have an assignment yet. Lately, to pay the bills, I've been working as a clerk at an insurance company."

"Well, I'm not a reporter at *Rolling Stone.* I compile the charts page." Around us, the group decided to chant the farmer's name in what they probably thought was a friendly, inclusive way, but he did not turn around.

Julie and I talked for an hour in our own bubble until the hay wagon rattled its way back to the farm. As everyone jumped down and dashed off to the next activity, Julie walked over to thank the driver, who was picking beer bottles out of the piles of hay, and I knew that my instincts about her were correct. Julie was missing the hard edge that afflicted so many of my city sisters and brethren. Julie, I would soon find out, was the type of person who wouldn't feel the need to comment on the lopsided wig that a diner waitress wore, who refrained from ordering in food during bad weather because she didn't want the delivery man to have to ride his bike in the rain. Tourists constantly asked her for directions, old ladies flapped over to her in the grocery line to compare purchases.

She told me that she once attended her superintendent's Tupperware party because she saw that he was inviting other people in her building and thought that nobody would come. "I was right," she said as we walked toward the farm. "Four people showed up, and one was his sister. There was a deli platter and about twelve bottles of wine for the five of us. And the super is a recovering alcoholic, so he doesn't drink." She sighed. "I ended up buying a hundred and twenty dollars' worth of Tupperware." As we continued to walk, talking intently, I learned that she had an encyclopedic knowledge of the history of baseball, gritty films of the seventies, and Halloween collectibles.

She shielded her eyes from the sun, gazing in the direction of the barn, where the family matriarch was sweeping the floor in preparation for the square dance.

"I have to find my parents," she said. "They'll want to be getting back soon."

My heart sank. "You're not going to the square dance?" I asked.

"No," she said. "I guess you're staying here?"

I nodded. "I'm in that big cabin," I said, pointing to a ramshackle building near the farmhouse.

She looked appalled. "The one with the burst pipe? I heard all about it on the hayride. How many people are in there?"

"Fourteen," I said. "Let me show you my horror." No one was in the cabin. Julie gingerly stepped in. Her foot made a squishing sound because the carpet was flooded with sewage from the burst pipe. Empty beer cans littered the living room. A rustic plaque of a Pa Kettle type hung on the wall, inscribed with the words, "I'd have to git better just to die."

"This is where I'm sleeping tonight," I said, pointing to a lumpy plaid sofa that smelled of long-ago ass. I bit back the impulse to ask her if I could stay with her at her parents' house.

She surveyed the room with her hands on her hips. "I am aghast at these conditions," she said. "I'm not a backpack-through-Europe type of person, let's put it that way. My idea of camping is a hot dog at Riverside Park. But this . . ." She held up her hands. "This is an outrage." She checked her watch. "I'm sorry to do this to you, but I really have to find my parents."

"I'll help you," I said quickly. Outside, she made her way to a couple who looked as uncomfortable as we did.

"Ready to go?" her mom said.

Julie gave me a note. "Here's my number," she said. "Call me anytime." She put both hands on my shoulders. "You can get through this," she said quietly.

She called me the day I got back to make lunch plans. "Quick," she said. "Turn to channel two. There's a squirrel licking a lollipop." I fumbled for the remote control.

The newscaster gave a hearty chuckle on the voice-over. "How many licks will it take for this little fellow to finish?"

"So where should we go?" she asked. "I warn you, I like to go to places that have been around forever."

"I do, too," I said. Not that I had actually been to any places that had been around forever, aside from rock clubs that only smelled like they had. I heard a flushing noise.

"I had to flush the toilet because I was cleaning my hairbrush and I just put in a big wad of hair," Julie said. "I wasn't going to the bathroom or anything."

"I understand," I said.

We soon made a habit of visiting the least happening spots in the city. Julie, already living in Manhattan for years, had been everywhere in town but had an affinity for quirky places from a gentler time. Our first meeting was at Rumplemeyer's, an ancient ice-cream parlor on Central Park South that had pink walls and a long counter with a soda fountain. There were stuffed animals for sale, and candy, and the whole place breathed a faint but reassuringly musty scent. We sat at one of the little round tables in the back and had ice-cream sundaes.

The following week she took me to Kaplan's at the Delmonico, a deli that had been around for nearly a century. "My father used to bring me here in the seventies," she said, pointing out the old-fashioned deli counter with yellow lights, the display case filled with Dr. Brown's soda. "All the waitresses call you 'sweetie.' " It was fake-wood-paneling heaven, and when you couldn't finish all of your colossal pastrami on rye, the waitress would wrap the pile of meat up for you and tuck in some extra bread so you could have a whole new sandwich later on.

Then we would walk to one of the small, chronically underloved museums in the city: the Merchant House Museum on the Lower East Side, a perfectly intact nineteenth-century house from a Bygone Era that typically had exhibits of Victorian mourning jewelry made out of human hair, or the

Abigail Adams Smith house, where workers gave tours in period costumes. To the world, Julie and I coolly displayed all the trappings of hipsterhood, but around each other we let our geek flag fly.

We called each other four times a day. We developed a shtick: If one of us picked up the phone, the other began talking as if we had just been in mid-conversation.

"Hello?"

"I just took a cab to work, and when we stopped at a light, my cab-driver opened the window and poured out a cup of urine," I said. "Is this common?"

"What? No. I would say that it isn't."

"What are you doing?"

"Deciding if I'm going to go to Alan's party." Alan was a fortyish typist at the insurance company where Julie worked who had a heavy Brooklyn accent and a wet-look hairpiece. There was a fussy dignity to Alan. He wore a smock because he didn't want to get the typewriter ink on his polyester suit. Once when I visited Julie at her office, Alan approached her desk, holding out an open box to us.

"Care for a Vienna Finger?" he asked.

Alan was known around the office for his holiday-themed parties at the Bay Ridge apartment that he shared with his elderly mother. His latest was in honor of Halloween, but sometimes he changed it up and threw an Autumn Party.

"You wouldn't want to come with me, would you?" Julie said.

"What, are you kidding me?" I said. "I'm in."

That Saturday I took my first trip to Brooklyn when we rode the subway out to Bay Ridge. Alan met us at the door, throwing it open the moment that Julie knocked. "Thank you so much for coming," he said in his decorous way. The entire apartment, from the shag carpeting to the walls, was pink, which nicely offset Alan's sizable collection of porcelain clowns.

"This is for you," said Julie, handing him a box of Godiva chocolates.

"Thank you," said Alan. "I appreciate it."

We walked into the living room, which was decorated with an explosion of plastic Halloween gewgaws and orange and black crepe paper. A few of Alan's neighbors introduced themselves and we chatted with two receptionists from the office who had come. "Where's his mother?" I asked quietly.

"She's stashed away somewhere," said Julie. "It's like when you're in seventh grade and your parents stay upstairs and you have a party in the basement." We made our way to a table that was covered with foil trays of Italian-American specialties. On the subway over, Julie had briefed me on the menu. "Part of the party ritual is to rave about Alan's cooking," she had said. "It's basically old-man Italian food. The big thing is *fritto misto,* which is battered, deep-fried vegetables. It's like having some broccoli in a doughnut. The other thing is rice balls."

Julie handed me a paper plate with a ghost on it. "Well," she said. "Let's dig in."

Alan came up behind us, bearing yet another tray heaped with chicken Parmesan, and slowly lowered it onto the table. "Alan," Julie said, holding up a hand. "Alan. These rice balls. Really delicious."

He puffed up. "I can't give out the recipe," he said.

"*Alan.*"

He shook his head. "I'm sorry." After the meal, Alan brought out the Godiva chocolates that Julie had given him and slowly walked around the room, displaying them to each guest, who would say, "Ooh, fancy" or "Don't they look delicious." Nobody actually got to taste them because Alan left the cellophane on the box.

"I used to work in catering," said Julie, watching Alan as he vanished into another room to stash the chocolates away. "There's an expression that they use called 'parading food.' Before everybody eats their salad, all the waiters parade around displaying the salad to everyone at the tables. It's a very big part of the caterer's oeuvre. I don't know why anyone would want to see their food before they've eaten it, but apparently some people like this parading of the food. Alan is one of those people. This is a chance for him to parade his food."

Indeed, the festivities peaked when Alan emerged from the kitchen holding a Halloween cake shaped like a jack-o'-lantern. Slowly, solemnly, he circled the room, exhibiting the cake to each person as though it were the Queen's jewels.

"Beautiful," I said when it lingered in front of me. Then he took out a camera and had us all surround the cake for a group photo. After that, Alan had a neighbor take a more serious solo shot of him alone, holding the cake.

Later, as we got our coats, I whispered to Julie, "I had the best time."

She nodded. "Me, too. It's almost like going to the Lower East Side tenement museum and seeing an actual family living there. You know? It's a slice of life you would never see. This isn't a goof. And it's endearing. It's touching to see the effort he went through."

She handed me my coat and smiled. "I couldn't bring just anyone, of course. I knew you would have the right spirit."

Booze: At Least As Important As Your Tape Recorder

If your subject is a reluctant interview, do everything in your power to get a drink into their hands. Alcohol liberates the tongue and blurs the time so that your allotted hour slips by unnoticed and stretches into six. Optimally, you should remain sober while your companion gets plastered, so as the evening progresses and your woozy new pal begins to spray your face with a light coating of spittle as he or she talks, surreptitiously switch to a mocktail. Around midnight, make a big show of "feeling dizzy" and wobble off to the bathroom, where you shut yourself in a stall and coolly take notes.

Only once did I deviate from my own advice, during a Lollapalooza tour stop in Atlanta. The bill was especially good that year: the Beastie Boys, the Breeders, L7, A Tribe Called Quest, Smashing Pumpkins, George Clinton and the P-Funk All-Stars. I flew down to interrogate the Breeders' Kim Deal for *Rolling Stone*'s special "Women in Rock" issue. I was to ask her the typically weighty questions that were posed to all participants: How has the role of women in rock changed over the last four decades? How are you affected by misogynistic lyrics in rock and hip-hop?

I met her backstage at the venue, where she sat on a battered couch in an oversized T-shirt and stained jeans, joking around with various crew members and musicians. As a former Pixie and member of the Breeders, Deal was one of my heroes. I loved the sound of her sweet, husky voice, and the way she smiled onstage as she played bass as though she was having the best time in the world.

Deal was perfectly friendly, but she was not in the mood to hold forth about being a victim of the patriarchy.

"So," I said nervously, fumbling with my notebook. "I know that these aren't the most freewheeling of questions, but maybe we can find a way to have a good time with them."

Silence.

"Do you feel like there is a glass ceiling in the music industry?" I began.

She groaned. Everyone around her laughed. I tried another question.

"What effect has being a woman had on your music?"

She rolled her eyes. "Do I have to answer this now?"

I considered. "I guess not," I said.

"Good," she said, producing a bottle of vodka and taking a swig. "Let's go see the Black Crowes. They're playing on the second stage." She held out the bottle. I glugged it down, figuring that I could get her to hold forth later, once she'd loosened up. After a few more lingering swigs, she jumped up, exhorting me to follow. We had to plunge into the crowd on our way to the smaller stage. As we waded through, a Gothy gang of teenage girls surrounded us, clamoring for autographs.

"I'm nobody," I told them. "My signature is worthless." They looked at me suspiciously, then, suspecting that I was simply being modest, shoved pens and paper at me with greater urgency. Feeling foolish, I signed as Deal smirked nearby. Then she'd had enough. "Come on!" she shouted, and dragged me back into the crowd. The Black Crowes were just tuning up. A few hands appeared out of the solid wall of fans in the audience, offering joints to us. Pot made me paranoid, tired, and hungry (three things I usually was, anyway), but, of course, I puffed away.

A publicist hurried over with beers for us, which stayed magically full, *Alice in Wonderland*–style, throughout the show. Joints! Beers! Crowes! Vodka! Joints! Beers! Crowes! Whoops, feeling a little dizzy. No, I'm fine, it's cool. Just going to crouch here for a sec.

"Let's go see the Beastie Boys," Deal shouted over the cheering crowd as the Crowes' set ended. She charged through the mob of fans with me in hot pursuit. She certainly seemed loose. I had to strike.

I paused on a stretch of lawn that ran between the stages, and shouted for her to stop. "Seriously," I pleaded, pulling out my tape recorder, "can you just answer a few questions? Only a few. Who were your musical heroes?"

"Later!" she said.

"Now!" I slurred. The lawn was lurching dangerously. *Storm's a-brewin', I reckon! Better bring her into port!*

"No!" she shouted, laughing crazily. Suddenly the ground shifted and *we were wrestling on the lawn of the Lakewood Amphitheater*. What made the scene even more surreal was that as we tumbled on the grass, a tall black man in diapers, a member of the P-Funk All-Stars, ambled past without giving us a second look.

"You will answer my questions!" I hollered, panting. I had no authority whatsoever. I begged, I threatened, I made jokes, but she wasn't having it. At some point, you just have to let it go.

Then, as the distant strains of the Beasties' "Sure Shot" started up, she wriggled free and broke for the stage where they were playing, urging me over her shoulder to follow. All doors were open to us as we headed backstage, then slipped off to the side of the stage, the Beasties mere feet from us. She pulled me next to the largest speakers I had ever seen and we danced through their entire electrifying set, to the amusement of a good portion of the audience, who had a full view of our rhythm-free flailing and leaping. *This is the best interview I've ever done,* I rejoiced.

Back in New York, my editor Karen studied my manuscript with a frown. "This is the worst interview you've ever done," she said. "What happened? It looks to me like she answered a handful of questions. Where's the rest of it?" I explained the saga of my struggle, conveniently leaving out the part about acting like a coal miner with a Friday-night paycheck.

"I'm sorry, but we'll have to kill this," Karen said, shaking her head. "There's nothing there." Even so, Kim Deal was a Woman Who Rocked. She just preferred to show me, rather than tell me.

9.

"Just hear me out," my friend Tina was telling me at lunch. "Go to the audition and if it's not for you, don't do it." Tina was an executive at MTV whose low-key manner belied her high-octane job. Being around her always made me feel like we were in a secret club, filled with intrigue and excitement. (A typical staccato phone message: *Yello, Tina calling. Here's the thing. I have a meeting at three, but how about we go shoe shopping at four. Check the schedule. Circle back.*)

As we split a piece of chocolate cake, Tina said that MTV was starting an all-music channel called M2 (later to change to MTV2). They were searching for a female on-air personality who had a decent knowledge of music. The list of New York–based lady rock journalists is a concise one, so my name inevitably came up. "I know you don't like doing television, but this could be fun," she said.

Sometimes, if I wrote a story for *Rolling Stone* that made some sort of splash, I would be called upon to do an interview for one of the celebrity news shows. I usually wriggled out of it, because I found the whole process excruciating. First, a crew would show up at the office and film you faking some activity so that filler footage could run while an announcer said, "Jancee Dunn is a reporter at *Rolling Stone.*" Usually the producer would instruct you to type at your computer, or chat on the phone to a nonexistent "source," or, the worst, stride purposefully down a hallway with a "stop the presses" expression on your face. This usually involves multiple takes because the producer will tell you that you're smirking, or look slightly demented, or that you're staring at the camera when you should act as if you're lost in thought, mentally writing your next story's lead paragraph as you walk along.

Then, for forty-five minutes, you dutifully answer a producer's detailed questions. That night, the extensive footage is narrowed down to one dopey, truncated micro-quote ("Sarah Michelle Gellar is a great girl, and—") before your answer is awkwardly cut off. Then they cue up the garish music and they're off to the Cannes Film Festival, where the stars light up the red carpet!

"I'm horrible on TV," I told Tina. "Trust me."

"Just give it a try," she said. "It's only part-time, so you can keep your job at *Rolling Stone.* I'm going to have a producer call you to set it up."

Two days later, the producer phoned to give me instructions for the audition, which was to take place, terrifyingly, on the street outside the MTV studio in Times Square. "Don't wear white, because depending on the lighting, it might glow," he said. "No patterns or stripes, because they can look animated, almost, on camera. You know how it can look like it's moving? And you'll probably have to read off of cue cards, but there won't be a lot written on them, so don't worry."

I worried. "How do I act? How do you want me to behave?"

"Just be yourself," he said. "Tell jokes, and if you have any information about the musicians you're talking about, throw it in there, because you want to seem informed. We'll send you a rough script, but feel free to say whatever you want."

The script introduced different videos with a couple of pertinent facts, such as new projects, tour dates, or background information about the video shoot. The channel only played music videos, and the variety was staggering. The playlist resembled some insane late-night cable access show:

a Buzzcocks video would run after Tiffany's "I Think We're Alone Now," followed by footage from an early-seventies James Brown performance. LeAnn Rimes, N.W.A., KISS, the Cocteau Twins, Nirvana, the Sugar Hill Gang—all were tossed in together.

For days, I hastily memorized trivia on the artists I was to mention and then practiced a way of casually throwing it out there as if I had just thought about it. *Oh, here's something, Gene Simmons was once an elementary school teacher.* On the day of the shoot, I met the producer on the street behind the studio.

"I'm going to hold up a cue card," he said, "but try not to look at it. Just talk to the camera as though you're telling something to a friend."

My lips were trembling. My smile was a ghastly, grinning skull. As I rattled out my fun facts, the cameraman swooped the camera around me to create a spinning effect. I assumed they wanted my delivery to be short and sharp, but I found myself rambling as I corrected myself, or mused aloud, or wandered off on tangents. This was supposed to be a hip new channel, and I sounded distressingly like Bob Newhart. Flop sweat beaded my forehead. They urged me to be animated, but my arms hung heavily at my sides. At one point, for variety, I clasped my hands together before they returned to their droopy job.

One script introduced a video from the Cranberries, so I put in a fact about the lead singer, Dolores O'Riordan. "She recently won a libel suit against a newspaper claiming she was cavorting onstage wearing no panties," I said. "Dolores claimed she was, in fact, wearing panties, and she won."

"That was great," said the producer brightly, afterward. "We'll let you know."

"Right," I said, quickly gathering my things.

A week later, a second producer called. "My name is Lou," he said. "I'm the supervising producer." He paused. "I saw your audition tape."

"I know, I know," I said. "Painful."

"I don't know how you did it," he said. "Apparently *somebody* likes you."

I got the gig. Taping took a half day, two days a week. My *Rolling Stone* editors tolerated my absence from the office, provided I got my work done. On my first day as a veejay I reported for duty at an East Village thrift shop. MTV2 didn't want a studio setting, so we shot at various locales downtown. We were allowed to wear our own clothes if we wanted, so I showed up with a carefully chosen green shirt with a subtle pattern on it. A slim, dark-haired guy ran over holding a cigarette in one hand and a clipboard in the other. "Hi, I'm Lou," he said, grabbing my hand absently. "It's nine hundred degrees in here." While the camera crew set up, we quickly got to know each other. In five minutes I extracted that he grew up in Hoboken, was an aficionado of bad made-for-TV movies that usually ended up on the Lifetime network, was a sugar fiend, and didn't have much of a problem speaking his mind.

He stared at my top. "What's with the shirt?" He pointed to what I thought was an artful splotch on my left breast. "One of your tits looks darker than the other. It looks like you're leaking milk." Before I could say anything, he pulled me over to a corner where a production assistant was setting up a craft services table with snacks, grabbed a bag of sour gummy peaches, and started to eat. "Get these away from me," he said. He continued to pop them in his mouth. Was I supposed to actually take them?

"Listen," he said, "we film dozens of segments that will run over the course of a few days, so you'll have to change clothes and hair." He introduced me to the wardrobe girl and a hair and makeup artist. "It's your new entourage," he said drily.

I changed in my "dressing room," a tiny bathroom in the back of the thrift shop. The wardrobe girl buttoned my blouse for me and tied a scarf around my waist as a belt. Then she stood back, squinting critically, then darted forward again, fussing, adjusting. After she nodded, I was passed to makeup. I was fighting to stay cool but my inner hillbilly kept bobbing up: *Garsh, there's free food on a table that you can jist go 'n' eat! And a lady who puts on your makeup! And I got me a lil' ol' intern gal who runs to git me a Diet Coke!*

Lou bustled up and told me to be ready in five minutes. "Oh, and the channel is on a satellite dish, so you need to say the transponder number."

"I'm too nervous," I said. "I won't remember. Plus, I don't know what a transponder is."

He handed me a piece of paper. "Well, read it off of this when you're on camera. We're very informal here."

The makeup artist slapped on some more powder and then we were ready. As I got into position and a sound guy attached a microphone to my shirt, another new veejay who had just finished her shift lingered in the doorway of the thrift shop.

"Lou," I whispered. "Why is she watching me?"

He raised an eyebrow. "She's not doing anything. I've got news for you, you're going to have to get used to people watching you."

I leaned in. "Yes, but this is my *first day.*"

He rolled his eyes and sighed loudly. "I'll try to get her into wardrobe or something." After he lured her away, he returned. "Try to relax," he said. "You look like you're about to have an embolism. It's just me and the crew, and believe me, we don't have any viewers. We're really working on our own *Private Idaho* here." I could feel some hives form on my neck. He peered at me. "What are those?"

"Hives," I whispered. "It happens when I get frightened."

He shook his head. "Can we get some makeup?" he yelled. The makeup artist stubbed out her cigarette and hurried over.

"Can you spackle her hives, please?" he said. She dabbed gingerly at the welts with a makeup sponge, keeping her face mercifully impassive. Then Lou turned to me. "Listen, if you mess up, try to correct yourself and just keep going. Okay?"

"Sure."

Then he counted down. "In five! Four!" Then he held up three fingers, two, one, and silently pointed my way. My lungs constricted as I stared back at the ring of faces—some bored, some interested—that surrounded me.

"Hello, viewers," I ventured. "You're watching the very first hour of MTV2. It's a new network with videos, videos, videos, twenty-four hours a day. I don't quite know what I'm going to be doing, but I do know they have me on board."

Lou nodded. Good.

I tried not to glance at the cue card that the production assistant was holding. "Coming up we have something new from Liz Phair and a classic from Bob Barley." I stopped, flustered, then I looked at Lou. "Can we do it again?" I asked, assuming he would stop the tape.

He shook his head.

"Please? I said 'Bob Barley.' I'm begging you."

"Keep going," he said firmly.

"Seriously, that was terrible—"

"No!" he said. "*I* will decide when we do it again! Keep going!"

Chastised, I continued. After a couple of takes, Lou pulled me aside.

"What made you decide to begin with 'Hello, viewers'?" he said, lighting up his twentieth cigarette of the day. "Who talks like that? Even Diane Sawyer doesn't say 'Hello, viewers.' "

"So do you want me to say something else?"

He shook his head while he exhaled a plume of smoke. "No."

Back we went for more segments. During one in which I talked about a clip from a tight-pantsed Billy Squier, who was clearly hanging to the right, I started to break down in a nerves-induced giggle fit. This tended to happen to me during somber events, like funerals. As I chokingly tried to Keep Going and deliver my lines, tears poured out of my eyes and I could barely croak out a word. Still, the camera continued, relentlessly, to film me.

At the next shoot, Lou informed me that the word from on high was that the network brass actually enjoyed my trainwreck delivery. ("Focus," said one letter from a viewer that Lou made me read aloud on the air. "Say your words. Pause and breathe. Take your time.") Because the channel had an organic feel—as if you had stumbled onto a cable access show programmed by a musical polymath—the slightly unhinged quality that I brought to the proceedings must have hewed to the Keeping It Real philosophy.

As I got more comfortable on camera, I began to love the job. The young crew worked hard but knew how to have fun. Every week we would set up somewhere different, and it was like a traveling party: a Broadway costume store, the basement of a sewing machine factory on the Lower East Side, a mouse-infested artificial flower factory, a dusty tenement on Twelfth Avenue, and once, memorably, a brownie factory. Taping at a bar could occasionally be hazardous. When the Deftones showed up at a sticky-floored downtown dive, lead singer Chino Moreno promptly turned a shade of garbage-bag green. "I just puked outside," he announced. "We were here drinking last night and the smell made me sick."

"I have Certs," I volunteered, reaching for my purse.

We filmed on the street, where homeless people would flap up to me and gibber away as I was trying to introduce a Black Flag video. The whole procedure had an appealingly lawless feel, particularly in the early days, when it seemed like our only viewers were in the Big House. We were constantly amazed at how many prisoners enjoyed full cable. Sometimes I would read their carefully composed letters on the air.

"Here is yet another letter from our friends in the Arizona State Prison in Yuma, Arizona," I would say brightly, holding a piece of tattered stationery aloft. " 'We love MTV2 and watch it for hours,' " I read. " 'Could you please play 'Man in the Box' by Alice in Chains, and also 'Freebird.' And please keep doing the mess-ups, they make us laugh, and we can use some laughs in here.' " I guess they thought my gaffes were scripted.

"No problem, fellas," I would say cheerily. "A video from Alice in Chains is coming up. And thanks so much for the sketch of the crying clown, I see some real artistic talent. Keep watching!"

Not only did viewers request videos, but crew members could, too. "Do you want to program an hour of videos?" Lou asked me one day. "The more obscure, the better."

"Really?" I asked. "You would let me do that?"

"We all can," he said. "Don't flatter yourself. I programmed four hours for my birthday. I put in every Kate Bush video there was."

So I programmed what I considered the most embarrassing videos ever made, among them "Hello" by Lionel Richie (plot: a gorgeous blind girl falls for art teacher Ritchie and fashions an elaborate clay bust of his head, each Jheri-curl on his clay mullet painstakingly detailed).

As the months rolled on and I adjusted to life on camera, my skin hardened to the consistency of a rhino's as my appearance was subjected to new levels of scrutiny—not from the executives at MTV, who, bless them, never said a single word about my looks, but from everybody else. If I would tape a show on the street, a crowd of bystanders would form who commented loudly on my appearance as though I were a hologram. "She's on TV?" one construction worker wondered as I talked to the camera. "With that ass?"

"Well," observed his companion, a guy whose paunch spilled over his cutoff shorts, "it looks like they're only filming her from the waist up."

I broke off in the middle of my spiel. "I can hear you, you know," I said to the guys, who both jumped as though I were a statue come to life. "I'm two yards away."

If I changed my hairstyle, mail would promptly arrive from my incarcerated pals, who commented more frequently than my girlfriends every time I lost weight or got new highlights. "I notice Miss Dunn that you cut your hair so that it is about one inch below your shoulders," read one dispatch from a Kansas penitentiary, written in stilted, warden-approved language. "I think that is a good length for you Miss Dunn but I liked you better with the darker hair but I do like the length as it makes your face look thinner."

I never realized until I appeared on camera that evidence of everything I did showed on my face or body. If I ate pizza the night before we taped, the wardrobe girl would announce, "Wow, you had carbs last night, huh?" as she was zipping up my pants (yes, she put on my pants). If I had been to a show and had a few drinks, our new makeup expert, an unflappable Brit who we called Sheree the Makeup Artist, would have to pull out the industrial-strength eye bag reducer.

As MTV2 grew, my encounters with walk-on guests would occasionally derail into embarrassment. I would often be so jittery before interviews that my skin would explode with monstrous stress zits. While magazine interviews were usually a one-on-one affair, TV was different. Usually it consisted of the artist, the crew, and the artist's sizable entourage of publicists, managers, and stylists, all of them gathered around the camera in a horseshoe, eyes trained, hawklike, on their star.

With magazine interviews, I could also control the outcome by excising all awkward moments from the text—my fumbling attempts to befriend, my inappropriately loud guffaws after a star's mild joke ("Hahahahaha! Whoo, that is some funny stuff!"). On camera, there was usually no time to make any sort of connection, fumbling or otherwise. The artist arrived, plopped into a chair already styled and readied by their squad back at the hotel, and we began.

During an interview with Outkast, I had just asked Andre and Big Boi my first question when the producer broke in. "Stop tape," she announced. "The zit on her cheek has skin flakes around it and the camera is picking it up. Does anyone have Scotch tape?" A production assistant was dispatched to find a roll as we all waited silently. Sometimes artists do not want to chat between takes, presumably to conserve their energy. I looked at Andre and gave him one of those tight, grimacelike, slightly downturned smiles reserved for coworkers you pass in the hallway at work, when you want to assume a pleasant, hail-fellow expression without actually chatting them up. Andre did not return the favor. Instead, he broke eye contact with me and tactfully gazed at the floor.

"Hold still," Sheree instructed as she ripped off the offending skin flakes while Andre, Big Boi, and their entourage sat quietly. My cheeks burned as I searched for something clever to say about skin flakes. Nothing turned up.

Worse was my encounter with Bono and Larry Mullen from U2, which they requested take place at an Irish bar in Midtown so that they could drink Guinness and play pool. The previous evening I had worked myself into a panic at the prospect of meeting one of my musical holies. I grew up with them, played their albums until they warped, pored over the liner notes as I lay on the floor of my bedroom. I even loved Bono's military-boots phase. God help me, I even loved that mullet.

The night before the interview, a three-dimensional pimple rose majestically out of my forehead like the Washington Monument. Alarmed, I speed-dialed Heather.

"I feel nauseous," I said.

"Let me guess," she said. "You have an interview coming up. Is it for *Rolling Stone* or MTV?"

"MTV, and I've got one of those hornlike pimples on my forehead." I flopped down on my bed.

"Oh, please," she said. "Why are you worrying? No one is even going to notice it except you. And it's just a celebrity. You've done a billion of these. Where are your Calms pills? Take a Calms. You'll be fine. Who is it this time?"

I paused. "Bono," I said.

She was silent. "Oh," she said finally.

I sat up. "Wait, why are you being so quiet?"

"Bono," she breathed. "Listen, did you try putting toothpaste on that zit? I heard that makes it go down. The paste, not the gel." She was quiet again. "If I met Bono, I don't even know if I'd be able to talk. He is probably the one person that would completely freak me out."

I took the phone into the bathroom and ransacked my cabinet, searching frantically for my Calms pills. Mama needs her Calms. Was the pharmacy still open? Heather was still talking.

"—I don't know if it would be better if he wore the sunglasses or not." She was musing softly, as if to herself. "Would it be better to look into the darkness or have to stare into his eyes? Probably the sunglasses would be less terrifying. Because if you look in his eyes you'd get locked, and then you would forget what you're talking about."

"You're not helping," I said.

"You'll be fine," she said quickly. "Remember: This is a huge event for you, but your interview is one of fourteen things he's checking off during the day. Plus, he has a reputation for being nice."

I showed up an hour early for the shoot. Lou hastened over. "I hate everyone today," he announced by way of greeting. "I walked to this shoot filled with rage. The whole fuckin' city is filled with twenty-five-year-old women talking talking talking on their cell phones, with their cigarettes and their lower back tattoos and their little fuckin' dogs. You know what I was doing while I was walking here? I was deliberately bumping into people on the sidewalk who were on their cell phones."

"Lou," I broke in. "I'm feeling a littl—"

"—I was just hoping they would say something to me so I could *lash out.* And you know what, I used to love Starbucks—like, instead of having a bag of candy in the afternoon I would have a mocha Frappuccino, but now I hate it, because every time I go in, there are these throngs of twenty five year-old women all saying that they need their Starbucks, and they're all in there slurping Starbucks like it's a *giant cock.*" He sighed loudly. "I just want to go home and make a list of people I hate, but I'm afraid that when I die, people will discover it and say that I was crazy."

His eyes flicked to my forehead and he abruptly stopped talking. "What is that?" he asked, frowning. "Did you hurt yourself?" He examined me from the side.. "I hate to tell you this, but it actually shows up if you're in profile," he said in a low voice. "Listen, when you're talking to Bono, it might be best if you faced the camera."

"But doesn't that look strange if he's answering one of my questions and I'm staring at the camera?"

Lou gave me a meaningful look.

"I look like a unicorn, don't I?" I said.

He shrugged. "It's a little distracting."

The band arrived with no entourage except for their publicist. Bono, at about five feet seven inches, was much shorter than I expected—you'll notice that he usually wears shoes or boots with a platform on them—but his charisma was like a crackling force field. He introduced himself to and shook hands with every member of the crew before making his way over to me.

Paul David Hewson! Born May 10, 1960, sleeps four hours a night, allergic to red wine, a chess whiz, owns two German shepherds! My mind whirred with arcane fan questions: Why have you never sung "Elvis Presley and America" in concert? Is it because you improvised the lyrics in one take? Would you ever put "Boy Girl," a track about bisexuality never released in the States, on a compilation? Did you really take off your clothes during an interview in a London restaurant in 1992 because you were bored? You mention in "Angel of Harlem" hearing something on BLS—is that the New York R & B radio station WBLS, by chance? Didn't you get the idea for "A Day Without Me" from a friend who tried to kill himself? Well, where is he now? Didn't Larry once get injections of bull's blood from a holistic doctor, and if so, what, exactly, was up with that? What did your wife, Ali, think of "The Sweetest Thing," allegedly written after you forgot her birthday during the *Joshua Tree* recording sessions?

I didn't have time to ask any of these questions as Bono smoothly commandeered the conversation. When he talks to you, he is utterly focused. In the ten minutes I was able to chat with him while the crew frantically rejiggered the lighting in that murky bar, the floor dropped away and the room spun around as we had a quiet, intense chat that touched on literature and politics and music and poetry. Well, mostly he talked, as I struggled to hold up my end.

As he was summoned to sit in front of the TV cameras, he made a self-deprecating joke that he had the face of a Welsh coal miner. "Oh Bono, that's not true," I tittered like a horse's ass. Lou, from his post behind the TV monitor, shook his head in disgust.

After I asked my questions, I stared straight ahead at the camera as they responded. *Don't turn to the side,* I kept telling myself. They must have been wondering why they had to address my ear, but they gamely went along with it, presumably figuring that I had some sort of showbiz tic.

As our interview wound up, Bono kissed my hand. I blushed, while Lou's eyes rolled so far into the back of his head that he could probably see his spinal cord.

It Can Be Done: Manipulating Your Way into the Kitchen Cabinet

If, despite your best efforts, you must meet your famous person at their production company or the office of their publicity firm, there are still methods to capture that all-important "color." Can you find a way to get a lift from the celebrity after your encounter? Cars can provide all kinds of notable details, particularly if your celebrity is messy. Stuck in a conference room? Try to get your celebrity to show you what is in their purse.

That said, when I heard that my chat with Dolly Parton was going to take place at her Nashville production company, I didn't worry. First of all, an American original like Dolly Parton does not need any extra "color." And I had a distinct feeling that her office would not be some barren cubicle warren. People with hard-core country roots cannot leave a bare surface be, and this is a woman who is drawn to all things that sparkle and gleam.

Even so, I knew that she kept a small apartment alongside the building, and I was determined to see it.

An assistant met me at the door of the low adobe building at nine a.m. sharp (Dolly, like Madonna, was known for being exceedingly punctual) and told me that Dolly would be along momentarily. The office was done up in a southwestern motif—lots of turquoise and peach, a cactus, a howling coyote statue. A life-sized cutout of Dolly in a tight red, white, and blue sparkly uniform stood by the entrance. The place was appealingly homey, so much so that the coffeepot was boiling over on a ledge and dripped languidly on the floor.

Her voice made itself known first—she was singing the old hymn "Peace in the Valley" as she tapped over in her five-inch stiletto heels. She stopped to greet me, put her hand jauntily on her hip, and patiently allowed me to gawk. She was wearing a spectacular platinum wig, a clingy black velvet two-piece pantsuit, big silver earrings, multiple shades of purple eye shadow, and shiny, bubble-gum pink lip gloss. "Well, hello," she said with a big grin. Who is like her, in all the world? Who is her successor? That mind-bogglingly small waist! Those glorious knockers! This glitzy getup was—hand to God—all for her trip to the chiropractor later in the afternoon. She cheerfully explained that most people would be frightened to look "this cheap and whorey," but not her.

I saw her take me in, too. Uh-oh, I thought. From her point of view, I must have looked like an uptight New Yorker—black clothes, preternaturally pale skin, reserved manner. Better establish my southern roots right off the top. But how could I do it in a way that wasn't completely obvious? Aha.

"I just ate the most delicious meat and threes at the Belle Meade Buffet," I told her, patting my stomach. "Cleaned my plate twice." I shook my head. "That's the kind of food I grew up with, and you just can't get it in New York."

She brightened, asking me if I was from the South.

I smiled. "Well, my mama is from Citronelle, Alabama." Suddenly she was my "mama." "We used to spend our summers there." I assumed a faraway look. "My pappy used to listen to your music on the radio when he was working in his toolshed." OK. Rein it in a little. It was true that I called him my pappy. His name was Hershal Ray Corners—where else could he be from but Alabama? But best not go overboard. Maybe later, I would name-check my aunt Eunice and my uncle Bud.

Dolly, sufficiently warmed up, came breathtakingly to life. She talked about her early years touring as a "girl singer" with the Porter Wagoner

Show before breaking to go solo, the controversy over her 1968 song "Just Because I'm a Woman," which lamented the double standard between men and women, and her myriad business ventures (beneath the folksiness, she is a sharply intelligent woman).

I was captivated by the staunch adaptability that had enabled Dolly to thrive after decades in the music industry. Each story was better than the last, but after a while I could barely pay attention, because I had drained the bottle of water she had offered me and, on top of the muy grande latte I had tossed back that morning, I really had to visit the facilities. Usually I never interrupt an interview for a bathroom break because it can waste a precious four minutes, but my concentration began to waver during a point when she was saying something especially interesting about her early days in a log cabin in Sevierville, Tennessee, when groundhogs and turtles and frogs were often on the dinner menu. Her father would lope off into the woods with a shotgun in order to feed his twelve children. My leg was jiggling. Think of something else. Think . . . of . . . something else.

"I have to use the facilities," I burst out miserably.

Not only did she shrug, but as I ran over to a nearby bathroom, she kept talking, even hollering through the door, just the way your girlfriend would if you were at her house. When I returned, she reminisced entertainingly about meeting her husband, Carl Dean, at the Wishy-Washy Laundromat in Nashville. Emboldened by her frank manner, I asked her if it bothered her that she had never had children. She was one of the few performers I had interviewed who did not, and since I myself had never been overwhelmed with maternal feeling, I wondered if she felt any sort of void.

She considered this for a moment. "I just think, you know, that it ain't meant for some people to have kids," she said. Instead, she felt like Carl was her child, and she was his. Besides, they took care of all of their nieces and nephews, sending them to college and buying them cars when they graduated.

Now that she had answered a question I was always curious about, I had to see that apartment. I wanted to see what was behind the curtain.

Behind the wig, if you will. So I concocted a strategy. I had read that she still snacked on Velveeta, as she had for decades. Maybe if I challenged her on it (it's fightin' words to accuse a country star of abandoning her roots), I could work my way into that apartment's kitchen.

"I heard that you still like Velveeta," I said, raising a skeptical eyebrow. "I have to say that I don't believe you. First of all, you have a child-sized waist. And secondly, at this point in your career you've got to have a personal chef."

She was indignant. "I do!" she hollered. "You want me to show you in my apartment next door? I fried up some SPAM yesterday morning!"

Yes!

She was up in a flash, racing purposefully through the office in those insane five-inch heels. She opened the door to a small apartment that was attached to the larger building. I was in! It smelled like incense, which was surprising for some reason. Racks of clothes lined the walls in an explosion of sequins, satin, and spangles. A dressing table was crammed with lotions and makeup. I wished I had time to inspect the clothes, but she kept going toward the kitchen.

It was cheery and comfortable, with blue Mexican tiles, notes and magnets on the fridge, and a Tupperware container on the counter filled with corn bread. Then Dolly triumphantly threw open the cabinets to show the most magnificent vista. There was a bomb shelter's worth of tinned SPAM, cans of corned-beef hash, loaves upon loaves of delightfully spongy white bread, and a giant brick of Velveeta.

Then, the pièce de résistance: She opened the fridge and fished out a ceramic pig. Inside was a bag of bacon grease, labeled with a date. It was a fancier version of my mother's ever-present jar of bacon grease in our cabinet at home. The *thwap* of bacon grease in a skillet—sweet music!—meant that we were having chicken-fried steak for dinner. Dolly, however, had her grease skimmed for her by the people who cleaned her house every Thursday. Dolly may have been worth three hundred million, but she was the absolute real deal. Who else could credibly say, as she did to me, "I have to have bacon grease in all of my houses"?

The trip to her kitchen was the high-cholesterol icing on the cake. She thoughtfully hacked me off a big hunk of Velveeta as a snack for my plane trip home. Then she pressed a bag of tomatoes in my hands that she had grown at the house. I carefully toted the bag home and ate every bit of the tomatoes. The Velveeta slab, meanwhile, stayed in the paper towel she had wrapped it in, which had a print of an old-fashioned country girl on it with a bonnet and calico dress. I just couldn't bear to throw it out. I proudly showed that Velveeta to the hotel porter, my cabdriver, and a flight attendant. How could I not?

"Well, of course she still eats Velveeta," said the hotel porter with proprietary authority. "Of course she does. Ms. Parton hasn't changed." A week later, my Velveeta was the same size and consistency, but I put it in the freezer, just in case, and there it remains.

10.

After a two-week period of flying to Los Angeles, London, and Paris for interviews (the best being the Spice Girls, which took place while all of us lounged on a bed in their cavernous suite at Hotel Le Bristol), I was feeling like a devil-may-care citizen of the world. This was a state of mind so foreign to me that I had to capitalize on it and hastily make the move out of my Hoboken apartment into Manhattan. The whole process of breaking the bonds to Jersey had to take place quickly before I reverted to my regular owlish persona, and what made it tougher was that Dinah and Heather had also moved to Hoboken after they graduated college. Their apartments were mere blocks away, so close that I would sometimes run over to see them in my pajamas. Dinah, the first of us to get married (to a genial chef named Patrick whose main passions in life were barbecue, the Giants, and Dinah), commuted into the city for her publishing job.

Heather, meanwhile, had also become a chef, having nicely parlayed her profound love of food into her livelihood. When she was on the job, she had met the man that she, too, would eventually marry. He was—to the everlasting joy of the family—yet another chef. If you couldn't work at Penney's, reasoned my food-obsessed family, this was the next best thing.

Heather's first job was at the Weehawken branch of the Chart House, a feel-good chain restaurant where the waiters wore Hawaiian shirts and a sign with a hook hung outside the employee entrance that said HANG YOUR BUMMER HERE. When Heather was promoted from the salad assembly line to the more coveted prime rib carving station, she developed a ferocious crush on a tall, handsome cook named Rob. Unfortunately, he was just as shy as she was, so they mostly communicated through group conversations.

Every night she would pop over from her apartment with a progress report. Finally, after a few suspenseful weeks, a breakthrough. "He talked to me," she said in a rush, bursting through the door and flopping dramatically on my couch. "He's so thoughtful and smart. He's going to cooking school and he wants to work in the city. He's so cute. His dad is from Puerto Rico and his mom is from Hawaii, so you can only imagine. He says he's going to give me his notes from class."

"That's a ruse," I said. "The old notes-from-class."

She leaped up, rigid. "You think so? Is it? No, seriously. Is it a ruse?"

The thought of leaving my sisters in Hoboken was depressing, but a friend of a friend was moving out of a cheap apartment in the West Village and said I could take over the lease. Of course, I called my folks to consult. Why did I do this? I didn't need their permission. "It's time I moved into the city," I announced to my father. A long silence followed.

"First of all, you'll have to pay an unincorporated business tax," he said at last. "And in terms of rental units, you're looking at a thirty percent increase in price. Maybe double, depending on the area. And right now you're paying a dollar for the PATH train, while the subway is a buck twenty-five, which means that annually it's an increase of . . ." I heard him fumbling around the junk drawer in the kitchen for his calculator.

"Dad, I already have a place in mind, and the rent is reasonable," I said.

"Those leases are full of hidden costs," he said.

After twenty minutes of heated negotiation, he accepted that I was determined to do it. "Get Polaroids of all of your valuables before you move," he advised. "And your mother and I are coming to see this place."

Exactly four days after I had moved in, my parents arrived—my mother holding a Tupperware container with a lemon cake inside, my father carrying his Stanley Jumbo Organizer Top Toolbox. Delicately, both stepped inside the door and took in the bleak surroundings: a five-hundred-square-foot studio painted the peculiarly dingy grayish white of all New York rentals. A few weak rays of light struggled through an airshaft window.

"Welcome, welcome," I said. They stood, frozen. "Sit down," I said heartily, pointing to a fatigued pullout couch. There was no room in the place for any other chairs. I glanced toward the bathroom. The john was a companionable two yards from the couch.

"Please," said my mother. "I'd rather you didn't sit on the toilet." She gingerly lowered herself onto the sofa, still holding the Tupperware on her lap, while my father inspected the windows. "The lock is broken," he said. "You'd better call the super."

"He's never around."

He crossed his arms. "Well, then, how about the owner of the building?"

"He lives in Israel," I said. "I don't know if he even speaks English."

My father raised an eyebrow. "Maybe he understands the words *Housing Court.* Maybe the words *New York City Citizen Service Center* will make him pay attention. How about that?" He stalked out the door. "I'm going to go find this super," he said. "He should be ashamed that he's not doing his job."

"Dad," I called after him. "Shame is not a motivator in New York."

Later, he returned, defeated, and the three of us spent a cheerless afternoon hanging pictures and exploring the neighborhood. When I walked them to the spot where their Buick was parked, I fought the urge to slip into the backseat and speed home to Jersey, as we had after our Bowery trips.

"I forgot to give you this," said my mother, handing me a clipping from the *Newark Star-Ledger* on starting an indoor herb garden and a small tub of peppermint foot cream. "This is my new favorite. It smells so good. Put it on before you go to sleep."

"I'm proud of you, kid," said my father, giving me a hug. "But start saving to buy an apartment. On a rental, you're just throwing a quarter of your income away."

I knew that on the drive home, my parents had a worried conversation about my suspended adolescence. I was nearing my late twenties, but because I lived in the city and worked in an industry that venerated youth, it was deceptively easy to travel in my own bubble of juvenilia, free of the quotidian markers of adulthood: mortgages and car payments and lawn maintenance and sitting on the bleachers on a Saturday morning, drinking coffee and chitchatting with the other parents as you watch your kid kick a soccer ball. I could barely conceive of the idea of getting married, let alone of having a child.

I barely had time to be lonely in my new apartment because the phone was constantly ringing with family members "just calling to check in." I had taken the day off to unpack and was balancing on a ladder, trying to clean out some high kitchen cabinets, when the first call came in.

I heard my mother's voice on the machine. "Hi, honey. Pick up! Pick up the phone. I know you're there. Pick up."

"I can't," I said out loud to the empty room. "I'll call you back."

"—Wait, your father is saying something. What? Jay, I can't hear you, you're mumbling. Your father says that maybe you're out getting that caulking that he talked to you about. Well, we're on our way to the spring flower show in Morristown and I thought I'd give you a call." My folks had recently retired, and they hit every craft expo, antique fair, and flower show in New Jersey with a vengeance. Without a job, my father reached new levels of preparedness, so their car contained bottled water, energy bars, hand wipes, a roadside emergency kit with flares, books on tape, a coin dispenser, and two hand towels to use as makeshift bibs so that they could drive and eat without making a mess. And a handheld recording device so that he could speak into it when lightning struck about fixing the garage door. "Jay, you turn left here," my mother was still saying. "I'm telling you. *Look at the sign.*

Fine, do what you want. Well, honestly, I think I know how to read a map, and it says Route Eighty, right here."

I sighed.

"So anyway, after this your father and I are going to see that movie about the detective. I forget the title. But it stars . . . who is that actress?" My father's disembodied voice, faint in the background. "Jay. Jay. It is *not Sally Field*. No, it is not. We just saw the pre—"

Beep. End of message.

I had just started scrubbing when I heard another message, this time from Dinah. "I've got something important to tell you," she said. "Call me, call me. I'm at the office."

My curiosity got to me and I climbed unsteadily down from the ladder. I took the phone into the bathroom so I could pluck my eyebrows at the same time.

I called her at the office and she picked up on the first ring.

"Dine," I said. I could hear some of her coworkers talking in the background.

"Oh, hi," she said hurriedly. "Listen, can I call you right back? I have people in my office and I have to run into a meeting in ten minutes."

I inspected my left eyebrow and plucked out a particularly stubborn hair. "Nope," I said. I loved to do that to her.

"But—"

"Now's really the best time for me," I said. I always said that. Then I would wait for her to demand why, exactly, my time was so important, but she never did.

"What? Well, okay." I heard her apologetically shoo some coworkers out of her office. ("It's an important call . . . right . . . I'll come and get you in a few minutes.")

"So," she said, a little thrill in her voice. "I'm pregnant. It's early, and I might lose it, but at least I know I can get pregnant. I'm due in February. I kind of hope it's a girl, but Patrick will take either."

"As opposed to you? What are you going to do if it's a boy? Leave it in a vacant lot?"

Dinah laughed. "No, no! I'll take anything, too."

My eyes brimmed with surprise tears. "Oh, Dine. That is the best news in the world." Dinah was the most traditional of all of us, and she had always hoped for a life filled with family birthday parties and vacations at the beach, just as we'd had.

"This weekend we're going to paint the baby room," she said. "It's too soon, but I want to do it anyway. Maybe in a green."

"Green's good," I said. The tears kept coming. Fortunately, she was called into her meeting and she had to go.

Dinah always left in a flurry of good-byes. "Okay, then!" she said. "Talk to you soon! Good to hear from you! Keep in touch! Miss you!"

I heaved myself off the bed and grabbed a tissue. Why was I feeling so melancholy? I had long told her that I wasn't a kid person, so it couldn't be envy. Maybe it was that Dinah's life had suddenly formed into a clear path. The whole family, including my folks, had their romantic lives locked up before they were of legal drinking age. I, meanwhile, didn't know what the weekend held, but wasn't that how I liked it? I had recently begun dating a p.r. hotshot who was witty and charming but wanted to keep things "loose," which meant a lot of last-minute plans. I could never call him right now.

Sniffling, I searched for one of the takeout menus that the last tenant had left, and ordered a grilled cheese from the kids' menu (it always had the best food) and a piece of chocolate cake. I was just splashing water on my face when the doorbell rang.

"How are you today?" the deliveryman said brightly. "Good! That will be ten dollars!" I handed him some cash and he bustled out. "Thank you, ma'am!" he called.

Ma'am? "I think you meant 'miss,' " I called back, but he was already gone.

Dinah's pregnancy seemed to touch off an epidemic of conception among my friends. The announcements arrived in waves from all over the tristate area.

"Come see the baby," said my high school friend Melissa, calling me one afternoon. Melissa was a fun-loving girl who drank her way through Boston College on a lacrosse scholarship. She did marketing for a hotel chain, commuting into the city from the suburban town of Summit. "Tyler's so lively now, you'll love him. I'll make some lunch. And bring Tracy."

Tracy was my closest friend from high school, a veteran from the days of Jersey Shore trips and basement parties and long afternoons of watching soap operas at my house while the other girls played sports. Now she was a stay-at-home mother of three daughters who lived in a large, immaculate house in Connecticut and ran a small catering business on the side. She hailed from a venerable old southern family in Augusta, Georgia, and knew how to wear pearls, and host twenty people for brunch, and write beautiful thank-you notes on creamy embossed stationery in her elegant hand. We lived through each other: If I was backstage at Ozzfest and felt myself cracking after my twelfth interview, I would call her to retreat into her genteel world of recipes and books and decorating, while she would phone me for celebrity gossip after a long day of hauling the kids to various lessons.

"Tracy," I said. "Please come with me to see Melissa's baby next weekend."

"He's four months old now, right?" said Tracy. She always knew the ages of every child. To me, they were small, medium, or large. "I just sent Melissa a selection of my favorite baby books. I love to do that. The second child gets a sterling silver picture frame. Sure, I'll go with you. Hold on." A small, reedy voice made a request in the background. "No. Elizabeth, we're having dinner soon. I'll be off the phone in two minutes. Stop bugging me. Go eat some M&M's."

"What's for dinner?" I asked eagerly. I had vowed, as a single girl, to make myself balanced dinners, but I usually ended up eating a random collection of unrelated foods: a bowl of cereal, a handful of baked potato chips, five olives.

"One of my favorite menus," she said. "Indonesian ginger chicken, which is luscious, curried couscous, and steamed haricots verts. And for dessert I'm making an apple galette. It's fabulous because you can use Pillsbury roll-up pie crust to make the dough, so it's a wonderful presentation without much effort." I heard the comforting rattle of pots as she prepared dinner. "I've got my book club meeting next week and I haven't even read the book because my parents are coming in from Augusta and I haven't had time," she said. "They're so worried about the snow here in Connecticut. My father wonders how the plane is going to land in the bad weather, while my mother is dragging out a pair of heavy boots that she, quote, 'once wore to Russia.' How about you? Whom have you talked to lately?"

I told her about my encounter with a serene, relentlessly positive Tony Bennett, which took place in the home office of his New York apartment. I owned fifteen of his albums and had applauded wildly at many of his shows, so I was apprehensive. I would have been crushed if he had been brusque, or phony, or distant, or creepy, or depressed.

"Come in, come in," he said, welcoming me with a beatific smile. He was drying a paintbrush with a cloth. "I was just painting. I usually wear a painting outfit and nice, cozy slippers." I immediately lowered and softened my voice to match his. He bid me to have a seat and I sank into a cream-colored wraparound couch. I relaxed instantly. He put a drink in my hand as music played softly in the background and the late-winter sky darkened to a lavender gray. A fluffy white dog jumped into my lap as snow began to fall, softly, gently, outside.

I petted the snoozing dog as Tony reminisced warmly about Irving Berlin and "the great Jimmy Durante" and his admiration for the crisp professionalism of Johnny Carson and Merv Griffin, who always wore ties and had great house bands on their shows. Tony's every other word was "terrific" or "fantastic." I wanted to spend the weekend drowsing on his couch. We could rent movies and play Scrabble, a bottle of Drambuie between us. After our interview, I told her, he sent me a bouquet of flowers, which I lovingly photographed.

"Well, he certainly sounds better than some of the hideous monsters that you interview," said Tracy. "All those bad-mannered men."

A week later, I met Tracy at the train station for our excursion to Melissa's house. Despite her long trip from Connecticut, her white shirt was free of wrinkles. Tracy always reminded me of a camellia: fresh and coolly pale. "I signed the card from the both of us," she said, carrying a large present decorated with ribbons the color of sugar almonds.

"You knew I'd forget to bring a gift," I said.

"Well . . . ," she said, waving her hand in a tactful way. "I know you're busy."

Melissa's husband, Daniel, picked us up at the station in his maroon minivan. Daniel was a good-natured banker with a flat Chicago accent and a propensity to sling his arm around whoever was near, a large, eager-to-please golden retriever, right down to the shedding.

"How is Melissa feeling?" asked Tracy. Melissa had apparently been in labor for thirty hours.

"Oh, she's holding up great." He laughed. "Tell you what, though, she was cursing me out in the delivery room. Tyler is a monster. A monster! He nearly split my wife in half!"

The traffic wasn't moving. "Let's go!" hollered Dan, pounding on his car horn, and the other cars jerked forward. Sometimes I love being with a giant, red-faced man.

I heard Tyler screaming as we pulled into the driveway. "Listen to the lungs on my little guy," Daniel said happily.

"Melissa's in the living room," he said, guiding us inside the house. "Wait until you see his hands," he announced. "Huge hands, huge feet. You've got to feel his grip." Every single new father in the Western world feels compelled to talk about his newborn's mighty grip and, if it's a male, his giant feet.

Melissa's sister, Sarah, grabbed my arm as I walked in the door. "Please don't freak out when you see her eyes," she murmured. "I know you get queasy." While Daniel disappeared to fire up the grill for barbecue, Sarah led me over to Melissa, who was wearing one of Daniel's tablecloth-sized T-shirts and smiling peacefully. The whites of her eyes were completely red.

"I know, it's bad," she said cheerfully. "I was straining so hard that I guess I burst some blood vessels in my eyes."

The volume of Tyler's hollering increased. "You got here right in time for Tyler's lunch," she said, peeling back the baby blanket. "You can't believe how often he wants to nurse." Oh yes I could. He was enormous and rubbery-looking.

"It's okay," Melissa crooned, lifting up her T-shirt.

I watched as her boobs flopped out of her nursing bra. They had enormous brown nipples. I had seen her breasts many a time in high school as she changed outfits before parties, and they looked nothing like that. These nipples looked like molasses spice cookies. My hand involuntarily headed to my mouth before Tracy shot me a warning look. I caught myself and forced it back down to my side.

Melissa caught me gaping. "Funny, isn't it?" she said, smiling wryly. "Some people think that the dark color is so that the baby can find your breast more easily, because they can't see so well."

I guess it is pretty funny when one of your body parts suddenly turns brown. Ha, ha!

She clamped Tyler onto one bosom and he began to gulp greedily, his translucent fingers pulsing softly like sea anemones.

Melissa's voice turned high and singsongy. "You're a hungry boy! Aren't ya! Aren't you a hungry, hungry boy! Yes, you are! Yes, you are! Sometimes you make my nipples bleed!" She grinned. "I can't tell you how your heart opens up when you have a child." Her eyes moistened with hormonal tears. "I just—I just can't explain it."

At that moment, Tyler's eyes swiveled toward me as he guzzled away. Then they narrowed. I tried for levity. "Don't worry," I said to him, waving my can of diet soda that Daniel had given me. "I've got plenty to drink." He continued to glare at me. Shouldn't he be tenderly gazing at his mother?

"Was it scary, being in labor so long?" I ventured.

Melissa smirked at me. "Look, you don't have to pretend like this is your thing. I know it isn't. But yes, by the end of it, I was begging the doctor to end my life." She shifted in her seat. "Ouch," she said. "You know, they cut me. Down there. I have eighteen stitches."

"And you have hemorrhoids," her sister prompted.

"Oh, yes," said Melissa, nodding vigorously. "Look out. Sometimes they don't go away for a year. It feels like you're constantly sitting on a pebble."

"Mine took six months to go away," said Sarah. They both looked at me expectantly. Was I supposed to share my own hemorrhoidal story?

Melissa brightened. "Hey, do you want to hold him?"

I must have looked alarmed because they both laughed. "Here," she said, detaching Tyler and handing him over.

"I don't know if this is a good idea," I began, but Melissa had already deposited him into my stiff arms.

"Be sure and support his neck because his head wobbles," she said. I pictured his head lolling, and then his neck snapping as I watched helplessly.

Tyler stared placidly at me. He was actually very cute, with rosy cheeks and a sweetly protruding upper lip.

"Take a sniff of his head," Melissa prompted. "It's the best baby smell." I stayed locked into position.

"Look at how stilted she is," she said to Tracy. I studied his face. Shouldn't I be aflow with tender feelings? And why was he sweating so much? It was as if he had just returned from the gym. Was that normal?

All three women were staring at me, so I felt like I should do something spontaneous to prove that I liked kids and wasn't a sociopath. I held him up over my head as I had seen people do in ads. He giggled and flapped his arms. It was sort of fun. I swung him up again.

"Blup," he gurgled, and then a cascade of milk flowed into my face, with a few drips landing in my open, smiling mouth. Human milk. My friend Melissa's milk. Melissa and Tracy scrambled for a tissue, laughing. "Whoops," said Melissa, doing what I thought was a very half-baked job of wiping my face.

I was still holding a squirming Tyler. Suddenly he went rigid. "Is everything okay?" I said, startled.

"Oh, sure," said Melissa. "He probably just has gas. He's really gassy. Aren't ya! Aren't you the gassiest boy!"

Tyler's eyes focused on a far-off point above my shoulder. Then he vibrated for a good five or six minutes. *Pffffffffffft.* Then he went slack. I handed him back to Melissa.

Dan came in the kitchen, clutching a barbecue fork, a driving rain of sweat soaking his T-shirt. "Need more marinade," he muttered.

I ran for their guest bathroom and Tracy followed.

"Forty-five minutes and we're out of here," I said, splashing my face with water. "I will never get the taste of that milk out of my mouth. It was sweet, and warm." I fought back a gag. "Spay me now."

Tracy passed me a guest towel. "Listen, I know Melissa isn't the person you know right now," she said. "But when you have a newborn, everything is just turned upside down." She put her hands on my shoulders. "I'm still the same. Okay? I had three children. Remember?"

I sighed. "I just don't get it. I don't know why you would want your house to explode with plastic toys. And did you see Melissa's eye bags?" I shuddered. "And why does every new mother love to tell you about the gory details? What was Sarah saying about a mucous plug? That was a new one to me."

She hesitated. "Well, it falls out of you before your water breaks," she said.

I opened Melissa's medicine cabinet and rummaged around. Hm. Daniel suffered from jock itch, it would seem. And somebody had a buildup of excess earwax. I picked up a long tube. "What do you suppose this is for?"

Tracy calmly took it from my hands and put it back.

I faced her. "Why did you have kids?" I asked. I had never asked her before. I guess I just assumed that she would give me the same answer that my mother did, which went along the lines of *I didn't think about it too much, I just did it.*

She smiled. "It sort of fit into my life plan," she said. "I felt like I was destined to get married, work, and then be a stay-at-home mom. I wasn't as career minded as you are and I'm very happy with that, settling into suburbia." She shut the medicine cabinet. "I'll be honest, you do lose a lot of brain cells, that's one reason for not having kids." She paused. "I've never considered myself to be one of those completely gushing over-the-top kinds of mothers, but without going all *Jerry Maguire* on you, there's a completeness to the picture. It's not for everyone. As Oprah says, being a stay-at-home mom is one of the hardest jobs in the world. But I really think you would be great at it."

"No, I would not."

"I wouldn't tell you that if I didn't mean it. You may not want to hear this, but you're incredibly old-fashioned." She shrugged. "You believe in your family so strongly, and you generally had a wonderful childhood, which is the cornerstone of being a good parent." I looked at her fondly. Tracy, whose taste ran to Lilly Pulitzer and Ann Taylor but never batted an eye during my heavy-eyeliner phase, who never drank anything stronger than a cosmopolitan but tried not to judge as I checked various drugs off of my to-do list. We trusted each other implicitly.

"Bur-gers!" we heard Daniel holler.

"Forty-five minutes and we're out," I whispered to Tracy.

"An hour," she countered as we pushed open the bathroom door.

Dirt-Gathering: Shortcuts to Finding the Least Loyal Person in an Entourage

If you are dispatched to a film or TV set and need to do a little sleuthing about your subject, forget the hair and makeup people. Yes, they love to gossip, but it's usually about everybody else, not the celebrity they are currently fussing over. Even if it's clear that they can't stand their charge, the most that you will get is eye rolling, because painting a famous person's face can yield a day rate of thousands of dollars, and working on a set means a steady paycheck, so it doesn't behoove them to tell you anything.

Ditto the crew. They are usually loyal to "the talent" because they won't want to be blackballed for future work. They want to remain firmly in the union, and who can blame them? And film sets are so incestuous that loose lips will swiftly be discovered. Production assistants and interns will flee from you in terror, chauffeurs often have to sign confidentiality agreements, and the catering people are never alone because someone is always hanging around the craft services table.

You need someone nonunion. You need someone who is completely mercenary.

You'll be needing the van driver.

Usually on a television or film set there is a scraggly guy who ferries various items or people around in a van. Often he is a local who does not work on sets for a living. He is your man.

I learned this unsavory fact when I was sent to Wilmington, North Carolina, to the set of *Dawson's Creek* for a rendezvous with then-rising star Katie Holmes. I had spent the day with her and found her to be sweetly wholesome. She told me she had grown up completely sheltered and happily naive in Toledo, Ohio, in the protective shadow of her older brothers. She attended an all-girl Catholic school, where the nuns told her that sex means love to girls and love means sex to boys.

In normal circumstances I would have liked her (although she had a grating habit of pronouncing "especially" *ex-specially*), but *Rolling Stone* wasn't *Ladies' Home Journal*. The cover images that sold the most briskly were of half-naked starlets, and we were encouraged to inject as much sex, drugs, and rock and roll into the text as we could reasonably get away with. Racy, she was not. I left our first interview with mixed feelings. Why did I have to tart everyone up? On the other hand, our chat had not been compelling.

Afterward, a bored *Dawson's* rep showed me around the set, then pointed at a brown van that was parked on a road near the entrance. "He'll take you to your hotel," he said.

A guy with a thin mustache and a tank top greeted me unsmilingly. He grunted, tossed his cigarette out the window, and turned on the ignition. Remembering my father's assertion that most unfriendly-looking people are actually shy, I cleared my throat.

"Not much action around here, huh?" I asked. He grinned.

"Nope," he said. "Most days I just sit around, have a smoke. It ain't so bad."

"How long have you been working here?" I said. After a while, I could ask questions on autopilot, and supply smooth follow-ups, without actually paying attention to what was said. It was all about modulation. Like a dog or a cat, I would snap into focus only if a voice abruptly raised or changed tone.

"About a month," he said. "I'm done next week. Got some business to take care of in Winston-Salem."

I nodded, gravely but sympathetically, as if I knew what he was talking about.

"Mind if I smoke?" he asked.

"No, go ahead." I hated cigarette smoke but wanted to appear breezy and hip.

He sparked one up and chuckled. "The only action I saw was with that little girl Katie and Josh." I sat up. That would be her costar, Joshua Jackson. Easy now. Clearly, he didn't know I was a reporter, but just assumed I was one of the many people who streamed in and out of the place. "They've been at it for a while, now."

"Oh?" I said lightly.

"It's the worst-kept secret on the set," he said. "Ask anybody." A Grinch-like grin crept across my face and I restrained myself from giggling. Thank you, my good man, and best of luck with your endeavor in Winston-Salem!

The next day as Katie and I had coffee together, I told her that I learned from a well-placed source that she and Jackson were an item. My conscience pricked me when her soft smile died and she buried her head in her hands. She whimpered a confirmation. I got my story. Everybody wins! Well, sort of.

In the music world, the tour-bus driver can be another rich trove of information, provided that he is a mercenary hire for a band's summer tour and not a regular driver with a decade of loyal employment. During a trip to Boston to spend time on the tour bus with the now-forgotten band Days of the New, I chatted up the bus driver while the group was outside arguing with their manager.

"Do you always work with these guys?" I asked.

"Oh, no," he said cheerfully. "It varies." Aha. I put my notebook and tape recorder down, signaling that I was off duty.

I smiled eagerly, like a fan. "Who was the wildest group that you've ever driven?" I asked breathlessly.

He considered for a moment. "Well, the strangest thing I've heard lately is about that fellow from Alice in Chains. A buddy of mine told me about it, he's a driver, too. That fellow—"

"Layne Staley?" I prompted. "The lead singer?"

He rubbed his chin. "Yeah. Him. That fellow does a lot of drugs."

"Right," I said.

"Well, my buddy saw him recently, and says that the guy shot up a lot, and sometimes he used dirty needles. Got gangrene in both wrists. Had to have both hands amputated."

My heart quickened as a familiar sensation flooded me—equal parts excitement and self-loathing at the discovery of a lurid story. "Are you sure about that?" I asked.

He swigged from his Styrofoam cup of coffee and allowed a small belch. "My buddy doesn't lie. He saw it with his own eyes." He looked at me pointedly. "When's the last time you saw a photo of him?"

I scanned my mental archives. "I don't know," I admitted. Staley hadn't been at any shows lately, but I assumed it was because of his drug jamboree, which was hardly a secret.

The bus driver shrugged. "That's because they want to keep it out of the papers. Although it's not like his career's over because he doesn't play guitar. I'm telling ya. The guy doesn't have any hands."

The next morning I flew back to New York and hastened to the office. I threw my bag down and speed-walked over to the cubicle of my editor, Karen. She was in the midst of closing a story and was staring intently at the screen.

"Listen," I said quietly. "I have it on good authority that Layne Staley *has no hands.*"

She squinted up at me as if I had a bug on my face. She had long ago become inured to most of my dramatic schemes but I could see she was curious. "What are you talking about?" she said. "I think I need a cigarette for this." She grabbed a pack of smokes. "Come downstairs with me."

I told her the story while she puffed furiously. Then she took me into the managing editor's office and made me repeat it. "Well," he said. "If it's true, then you'll have to do a story on it. Let me just check with photo to see if there is anything recent on him."

My heart leaped. Breaking news, and I was at the front lines! I trumpeted my report to a couple of the younger staffers, reveling in their satisfyingly shocked reaction, then walked purposefully back to my desk to commence my research.

Five minutes later, Karen dropped a photo of Layne Staley—both hands perfectly intact—on my desk.

"That was taken last week," she said. I could see she was holding back laughter, but she softened when she saw my crestfallen face. "You shouldn't be so disappointed to learn that a person has hands," she said. "It's good news, not bad news, that the poor guy isn't an amputee."

Then I started to laugh, too. Although I wasn't laughing a week later when coworkers were still greeting me by hiding a hand in their sleeve and waving a cheery hello with their "stump."

11.

I met Sean at an East Village roof party. He was standing off to the side of a knot of art-school types who were animatedly talking—the males in cutoff fatigue shorts and Chuck Taylors, the females with girly tattoos (fairies, a flower chain around the ankle) and hair adroitly piled on their head in that careless knot I could never quite approximate. Did they use bobby pins?

Sean was wearing a plain white T-shirt and jeans that were spattered with paint. Unlike everyone else at the party, he didn't look self-consciously hip. He just seemed vulnerable with his sad brown calf's eyes and tumble of honey-colored curls as he attempted to break into a conversation monopolized by a guy in a T-shirt that said Total Dick. To me, his skinny shoulders said "underfed" rather than "fashionably slinky." He needed to eat, but the only vittles were some soggy ruffled potato chips still in the bag. There was never any decent food at these hipster parties. I would have happily settled for some "ironic" onion dip.

He caught my eye and smiled shyly. *Save me,* he mouthed.

I walked right over to him.

We were still talking as the last guest left the roof. Sean was the only child of Vermont professors who enthusiastically supported his career as an illustrator, even after they paid for his degree in Eastern philosophy from Sarah Lawrence. Unlike the tightly strung p.r. executive I had recently dated, Sean didn't own a suit. He was gentle and shy, and he liked to cook and go on excursions upstate in one of his parents' castoff Volvos. He laughed at even my lamest jokes. I'm a sucker for a grown man who giggles—not a high-pitched serial-killer sort of giggle, but a lighthearted laugh.

He took my phone number, and the next night we went to dinner and the movies. Some of my friends complain that this sort of outing shows a distinct lack of imagination, but I find those top-of-the-Empire-State-Building dates too artificial. He brought a flashlight to the cinema, tucking it into a messenger bag. When a couple a few rows ahead of us started arguing loudly about whether the leading man had had some work done on his face, Sean walked over with his flashlight and shined it right into their eyes. "The movie has started," he said in a brisk, official tone. "Let's keep it down." The couple gaped at him, confused. Was he an employee?

He clicked off the flashlight and returned to his seat. "Works every time," he whispered.

Well. I was smitten. Sean had the kind of freewheeling personality that I envied. At home wherever he was, he rebuffed the New York eyes-forward rule and struck up conversations with any schmo in speaking range. Tattered old gals spotted his friendly face at fifty paces and sensed correctly that he wouldn't edge away from them but would happily chat away, sometimes for hours. When he started to make me dinners in my tiny apartment, they wouldn't be served until late, because we could never get out of a grocery store. He would end up trading recipes with some lady named Sylvia who had lipstick traveling up to her cheek and a Lord of the Onion Rings T-shirt.

With Sean, no activity was mundane. If we were on the subway, he didn't filter out the noise as I did but would troll the car for conversations that amused him. Once he nudged me and flicked his eyes in the direction of a gaggle of burly guys of various hues who were intently comparing the merits of cheese fries throughout the city.

"The ones at Brothers Bar-B-Cue have a lotta cheese," said one guy with a thicket of back hair poking out of his tank top. "A lotta cheese. You really get your money's worth."

"There were some good cheese fries at that Yankee game," his friend broke in. "And the fries had the skins on them."

Another guy waved his hand dismissively. "I don't ceh for that," he said. "I don't ceh for the skins."

"What about the ones we had at Rockaway beach, on the boardwalk?" put in a third. "That was some mad cheese."

"Those cheese fries were three dollars and fifty cents," said the tank-topped man indignantly. "Them shits was expensive. Yo, for three-fifty I could buy a pound of cheese and make my own damn cheese fries."

The debate raged on for fifteen station stops. "I love this town," said Sean as we stepped out onto the subway platform.

Sean told me I was his soul mate two weeks after we met. I felt that the polite response was *You're my soul mate, too.* I didn't know whether this was true, but I didn't mind when he promptly moved most of his stuff into my apartment. Isn't that always the way? Is it because women have clean towels? He was never eager to go back to his place in Brooklyn, which he shared with two roommates, one of whom wrote his initials on his particular third of the eggs.

Even though our space was cramped, I loved playing house. It was a tonic to come home from a trying interview to a sunny greeting and some new stir-fry concoction that he had made for dinner. Sean was an avowed vegetarian, and I admired that he had organized his life around his beliefs, while I was largely belief-free. He didn't eat "anything that had a face" and spent his weekends doing volunteer work for the Earth Society and the Waterwheel Alliance and some organization that worked to prevent "environmental racism," whatever that was. Most impressively, he didn't own a TV. "I'd rather experience life than watch it on a little box," he would say. He was horrified by my viewing habits. One night he clicked off the TV and brought me to Red Hook, in Brooklyn, a semidecrepit port where we peered inside old warehouses and climbed the rocks along the shore. "Wasn't that better than TV, the opiate of the masses?" he demanded as we rode the F train home.

"I thought that was religion," I said.

"That, too," he said.

There's something seductive about the whole crunchy lifestyle, with its scented oils and sexy yoga instructors. I soon fell into step alongside Sean. I may have been glitzy on the weekdays, but on the weekends, when I was shopping organic and writing checks to the Save the Manatee Club, my life was rich with meaning!

Sean's easy disposition made up for the fact that orders for his illustration work weren't exactly pouring in, but as he often told me, he didn't need money to have fun. He would laugh at my careful financial planning, the money charts and graphs that would arrive weekly from my father.

His favorite expressions were "No worries" and "It's up to you," which I found refreshing after Ritchie, who hijacked every plan and found my friends "boring." I thought Sean's willingness to hand over all decision-making to me was actually a confident move. He was happy to absorb all of my friends as his.

"What do you want to do tonight?" I asked one evening after we had finished a chick-pea-and-cumin concoction. "There are a couple of good movies at Film Forum."

"I don't care," he said. "It's up to you."

"Let's go to a movie, then," I said, grabbing my purse. "Let's get a big tub of popcorn and some Sno-Caps, and we'll sit in the air-conditioning and hold hands." I had picked up this habit of setting the stage from Heather, who liked to drum up advance excitement before any activity with a vivid picture. *Let's go to the mall, and we'll sit at the counter at the Neiman Marcus café and have their Confucius chicken salad and iced tea, and then we'll go buy lip gloss, and try on shoes, and then we'll walk around the bookstore, oh and then we'll get ice-cream cones. Maybe Rocky Road.* Or, if you'd had a hard day: *I know what you should do. Go buy a big pile of guilty-pleasure magazines and some slice-and-bake cookies, then turn off the phone, take a bath, and get in your pj's. Camp out on the couch with a cup of peppermint tea and a big fat blanket and a couple of pillows so you'll be all cozy. Turn on some soap operas, put the cat in your lap, and then just enjoy yourself and flip through magazines and have some nice, warm cookies.*

In a short while I thought of the apartment as Sean's. Somehow he had managed to physically coat the place with his presence. Even when he wasn't around, little piles of organic souvenirs abounded: loamy mounds of discarded socks that still held his foot shape, Fritos-like fingernail clippings, crumpled napkins with a smeary mouth print of lentil soup or vegetarian chili. A light sprinkling of his cast-off chest hair seemed to blanket every object in the place, and he wasn't even particularly hirsute. He had a habit of lounging around the place nude, which I would have found annoying with anyone else. Instead, I would just shake my head and laugh—proof, I felt, that I was adopting Sean's freewheeling ways.

After a few months, I felt it was time for him to meet the family. I started by inviting Dinah and Patrick to dinner.

"Why do I have to meet them?" Sean pouted. "I can meet them later."

"You know from all of the phone calls I get that we're very close."

He came over and gave me a hug. "I'd rather have you all to myself," he said. Was that endearing? Or creepy?

"It's just one night," I said.

That Friday, Dinah and Patrick burst in the door in their usual typhoon. She was four months pregnant but had as much energy as ever. "Hi, Sean," she said, giving him a kiss.

"Hi," he said, hugging her woodenly.

"Hey, buddy," said Patrick heartily, vigorously shaking his hand. Patrick was the sort of burly, reliable, self-deprecating guy that most men liked on sight—the one presiding over the barbecue that the kids hang all over because he's not afraid to be silly—but Sean hung back.

"Cute," Dinah whispered when he went to the kitchen to check on the tofu Mee Grob that he had made. "He looks like that singer from INXS." When he returned, Patrick was standing by the window to cool off. A self-described "large, sweaty man," he was frequently seen at family parties out on the deck, swabbing his face with his ever-present bandanna. Dinah, meanwhile, was unpacking a potted begonia and a candle for me. She always brought gifts. "It's more money than it needed to be, but I know you love candles," she said.

Sean examined it. "You could get one of those religious candles in the supermarket and save yourself thirty dollars," he said.

Dinah started for just a second. "Right," she said brightly. "You're right." Where was the perpetually smiling Sean, the one who did funny imitations?

At dinner, Sean swatted away Patrick's attempts at sports talk and Dinah's stab at politics. I steered the conversation toward travel. "Sean and I are going camping," I announced.

He brightened. "We're headed for Utah next month," he said. "Zion National Park. Eight days in a tent. It's going to be awesome. Also, I've worked it out so that if we eat protein bars for breakfast and lunch, we'll only spend a few hundred dollars for the whole trip." I could feel Dinah's eyes singeing the side of my head. I knew what she was thinking. I was the one who made her throw out her old sheets for something with a higher thread count, and urged her to buy pricey "investment" shoes. But Sean had sold the idea so well: nights under a blanket of stars, skillet corn bread in the morning as the sky grew pink. Sean had said that he wasn't going to bathe the whole time, just to see if he could do it.

Sean jumped up. "Anybody hungry?" He brought back the Mee Grob and plopped it ceremoniously onto everyone's plate, then passed a big salad festooned with sunflower seeds and carrots and raisins.

"Wow," said Patrick, carefully examining his dish. "What's in here? I'm a chef, you know, so I love trying new things." I could tell he was going to stop for pizza on the way home.

"It's a Thai dish," Sean explained. "It's just rice noodles with some tomato paste, and shallots, and cilantro. Stuff like that. And tofu, because I don't eat meat."

Patrick looked up, alarmed. "Nothing? Steak, chicken, pork, nothing?"

Sean shook his head. "And I'm trying to eliminate eggs, although it's hard. But it's not like you need meat to live."

This was like tossing chum in the water to Patrick, the barbecue enthusiast. "I understand where you're coming from," he said slowly. "But I always kind of believe that we're at the top of the food chain. To be honest, what purpose do chickens serve?"

Sean shook his head. "There are so many great man-made foods that you really don't need to eat meat, okay?" he said, raising his voice. "Soy products can taste like anything you want."

Patrick was barely listening. "God. I would be so depressed without meat," he said, his eyes staring blankly. "It's like a . . ." He searched for a properly bleak analogy. "It's like a . . . like a world without football. Maybe I'm being dramatic, but to me, it's like a world without loved ones." He put his fork down. "It's everything that's associated with meat," he said feelingly. "It's family and summer barbecues, and what would Thanksgiving be without a turkey? Imagine you're going to someone's house for Thanksgiving. You're sitting around, you've watched some football, and your relatives and friends are there and they say, 'Dinner's ready, let's come break bread,' and you're all sitting around the table. What's the moment you're waiting for?" He looked meaningfully around the table. "For ninety-nine percent of the people in this country, it's the turkey. It's the golden brown turkey that's been roasting in the oven for eight hours. It's, it's, it's the *thrill* and the anticipation of how it's going to *taste,* and the fun you're going to have tearing into this turkey."

He paused. "Now imagine the same scenario," he said darkly, "and on that same platter where the turkey should be, there's a head of *cabbage.*" He took out his bandanna. Clearly, he was worked up. "You'd think, 'Fuck, you've gotta be kidding me.' Cabbage? Even if you eat turkey once a year, the turkey's the centerpiece. Do you know what I mean?"

Dinah and I gaped at him admiringly, but Patrick, exhausted, didn't notice. He had a brow to mop.

From there, the evening slid speedily downhill and they left as soon as they could.

I turned to Sean immediately after shutting the door. "Why were you acting that way?" I demanded.

He shrugged. "I just don't see why I have to be involved with your family. I don't ask you to meet mine."

"But I would. I would, gladly."

He tried to hug me, his way of warding off arguments. "Relax," he said.

I broke away. "That is so passive aggressive," I said irritably. "You constantly tell me to relax. Here's a news flash: Women hate being told to relax."

"You're working yourself up," he soothed.

"What's so bad about working myself up?" I hollered.

"Shhh," he said, trying again to hug me.

"Goddammit!" I shouted. "Women hate to be shushed!"

After an hour of doing the dishes in silence, Sean jollied me back into a good mood.

"I'm sorry," he said, giving me one of his twinkly smiles. "Why don't we invite Heather over? Hmm? And your parents, too." We started cleaning up the dishes. "Hey, don't throw out that Baggie," he said.

"But it has schmutz in it," I said.

He shook his head. "It's called 'sustainable living,' " he said, grabbing it and rinsing it out. "Earth Day is every day."

I used to admiringly repeat his remarks to Heather. I would now use a stoner voice when I imitated him.

The Unfortunate Remark: A Moral Dilemma

On rare occasions, celebrities will veer from their carefully bland, publicist-approved sound bites and make a blundering comment that exposes them as vapid or foolish. If this happens, do not examine your conscience. Print their transgression, and while you're at it, blow up that particular remark and use it as the pull quote in your story.

Would that I could follow my own advice. I just never had that killer instinct. I see the way that one horrible quote can live on in infamy, dogging a person until they are onto their third face-lift. So I cave. Consider this forehead-slapping self-obsessed quote from a young blond actress who is the object of obsession for millions of preteen girls.

> *"I was in a movie theater the other day. And this little boy was choking on, like, a candy in the middle of the movie theater. And, like, I don't know CPR. It's just, like, it happened in a movie theater, where it's just like—you can't like enjoy the rest of the movie."*

I couldn't print it. I should have.

How about this horrifyingly dopey gem from a handsome, drug-friendly alternative rocker?

"I lived right down the street from Ground Zero. It was definitely a shame that people died and shit."

Dude, it was definitely fully a shame! I didn't have the heart to use it. It was just too embarrassing, and he wasn't an evil person, just dense. I have always been uncomfortable with that kind of power to manipulate, because the truth is that you could interview someone from the Brookings Institution in possession of five degrees and still make him or her look like a stooge. All you do is throw in all the "ums" and "uhs" and include the preinterview banal chitchat from the getting-to-know-you period before the tape recorder is turned on, time which is still perfectly fair to use.

That's my sad rationale for not "going for it." I just can't. But if you find yourself in a similar position, you should.

12.

The next morning I called Dinah for the postmortem on the Sean dinner. "Dine," I said. "Can you do me a favor?"

"Sure," she said cheerily. She really was so like my father.

"Why do you say 'sure' before you know what I'm going to ask? You always do that. Then you're trapped. What if I'm asking for something major? You should stall and say 'What is it?' Or 'Let me check my book,' or 'I'm pretty busy, but I think so.' "

She laughed. "But I would do you a favor."

I sighed. "Doesn't it bother you that I don't say 'sure' when you ask me for favors?"

She thought for a moment. "Not really," she said. Dinah was that sunny. She assumed that all people were good, and telling the truth, and had honorable intentions unless there was stark evidence to the contrary. I often accused her of living in Disneyland. Her response? "I like living in Disneyland." I ask you: What do you say to that? I once took her to Minneapolis to keep me company while I interviewed Pat Benatar at a tour stop. Dinah was right at home in the Midwest. At one point we went to a drive-through for burgers.

"How are you!" cried the fast-food employee, an older guy with ruddy cheeks and a wide smile.

"I'm doing just fine!" Dinah exclaimed. "We're visiting here from New York! Well, actually, I live in New Jersey, she lives in New York!"

"He doesn't care," I muttered.

"Well, isn't that great!" shrieked the employee. "How are you liking it here?"

I shifted in my seat. People were queued up behind us, but of course nobody honked.

"People are so nice!" Dinah caroled. "We just can't believe it!"

The employee nodded good-humoredly. "I betcha you girls are going to the Mall of America!"

I poked Dinah. "Give him the money," I said.

She chuckled. "We're going to the mall right after this!"

"Well!" he cried. "Isn't that going to be fun!" Back and forth they went as my fries calcified inside the bag that the employee held motionless above the car window.

There was no changing Dinah, so I abandoned my attempts at making her more guarded.

"Listen," I said. "Tell me what you really thought of Sean."

She hesitated. "Well, what do you think of him?"

"He says I'm his soul mate."

"Yuck," she blurted out. "Come on. You're not his soul mate."

"No."

"He's not my type of guy, but he seems nice," she said. "I don't know. You don't seem like you have anything in common. And it looks to me like you have overhauled your whole life for him, but he doesn't appear to be bending for you. Come on, you hate camping. What do you want from him, exactly? What can you learn from him?"

I thought for a moment. "Well, he's made me more laid-back."

"But you're not laid-back. You never have been. And it's *okay* not to be laid-back. You live in New York." She sighed. "I know you've never been promarriage, but—"

"It's not that I'm against it," I interrupted. "I'm just afraid of losing myself. I'm almost thirty and I feel like I'm just getting started."

She tried for lightness. "Patrick got pretty emotional about meat, didn't he?"

I laughed. "It was like Roosevelt's 'Four Freedoms' speech, only he tacked on a fifth one."

"I mean, I can almost be a vegetarian for the logical reasons," she said. "But Sean didn't have a reason for his view. Some people just do it because it's another lifestyle choice, another uniform. It seems to me that he's just doing it because it's cool. And what's with the stinginess? He seems willing to spend money, as long as it's yours." I knew I shouldn't have told her about that expensive Pinot Grigio that he had ordered on *Rolling Stone*'s tab at dinner last week.

She paused. "Look, he's very cute, and it's clear that he adores you," she said.

That wasn't the most ringing endorsement, but I took it, and hurried her off the phone. I was the only one in my family who was single. My dating travails were a source of snorting amusement for my parents, who met each new "creative type" I brought home with polite reserve and arch comments about grooming and presentation. They were clearly holding out hope that I would meet some fraternity captain with a firm handshake and a bright future as an options analyst. Well, that was not going to happen.

After my conversation with Dinah, I found to my extreme irritation that her comments made me view Sean with more narrowed eyes. I had always been slightly impatient, but tolerant, of his foibles, but it curdled into full-blown aggravation one night after I came home from a long day of five phone interviews to find Sean scrambling up groggily from the couch. He was nude, of course. A worn copy of *Zen and the Art of Motorcycle Maintenance* lay on the coffee table alongside a moldering hummus sandwich. "Hey," he said sweetly, giving me a too-tight hug. He laughed and smelled his pits. "Oops," he said with a grin. "Sorry about that." He bent over a pile of CDs, rummaging through them with his nut sack dangling like a turkey wattle. Then he shuffled off to the bathroom.

Damn you, Dinah.

Sean was oblivious to my rapidly curdling feelings. I began to believe the old adage that the qualities that initially attract will eventually repel. In my mind, "freewheeling" would soon change to "freeloading," "groovy" would quickly become "gamy."

"Hey, babe?" Sean called through the open door of the bathroom. "Can you bring me some more toilet paper?"

"Sure," I said. I tossed a roll through the door and waited for a moment. "Sean," I ventured. "I've noticed you haven't had that many illustrating jobs lately."

I heard the sound of flushing. "No, but things will pick up," he said cheerfully. "No worries."

No worries. "Sean, what the hell are you talking about?" I snapped. "No worries? Ever? Worries don't exist?"

He emerged from the bathroom. "Hey, hey, hey," he said, reaching for me. Did he wash his hands? I didn't hear the sink running.

As the days wore drearily on and his charms completely evaporated, I knew I had to get him out. If only he weren't so pleasant. To amuse myself while I devised an escape plan, I wrote haikus in my head every time he did something that thrummed my nerves. Which was often.

Empty milk carton
Remaining for weeks in fridge
Like Lenin entombed

Blue sponge for bathroom
Yellow is for the kitchen
Do not interchange

Sweater made of wool
Does not go in the laundry
Oh well, too late now

"Sorry about the roaches," said Sean. "I was just trying to make compost."

My toothbrush is pink
Is this hard to remember
Your toothbrush is blue

"You're not doing him any favors," said Julie. "Get it over with."

Is today the day
Garbage truck driving away
Try again next week

Pooper scoop unused
Nowhere for the cat to go
Except for the rug

"You are missing out on the best season of *Oprah,* ever," said Heather. "I'm just saying."

Do you have the keys
Are you sure you have the keys
I'll call the super

Beck CD cover
Yet Gang of Four is inside
Where's the Beck CD

"Jesus Christ," said Lou. "I can't hear about this anymore. Just get rid of him."

Ancient bag of trash
Is not a green beanbag chair
Or an ottoman

"What is it?" Sean said, his eyes anxious. My heart contracted. He really did not know. "What did you want to talk about?" He sat down on the couch. "Why do you look so serious?"

It's Beluga, Baby: How to Get Over a Boyfriend with the Help of the Ultimate Ladies' Man

If it is done in the right way, exposure to stars can have a rejuvenating, even a therapeutic, effect. If you are feeling low, try to remember that your rendezvous with a celebrity does not need to be a nail-biting experience. Time with a celebrity can be well spent. They can make you laugh (particularly if it is David Spade, who had me giggling continuously for two days straight as we drove around his hometown of Phoenix). They like to convene in glamorous locales. And if a celebrity flirts with you, it is an undeniable mood-lifter.

This was the case when, soon after a breakup, I dragged myself to an interview with Barry White, and thank the good Lord that I did.

My recovery began the moment I arrived at White's cavernous suite at his usual spot, the Helmsley Palace in New York, and surrendered myself to his tender care. Barry White—and I say this unnecessarily—knew exactly how to treat a lady. He greeted me with a friendly, lingering kiss on the mouth, despite our never having met. As he swooped in, his crunchy goatee tickled my chin. He was wearing the largest black leather vest I had ever seen, and he was casually snacking on a bag of M&M's—the big kind that you get at the drugstore to share with your office-mates. His own music was throbbing in the background at a discreet, sensual level. "Come on in, baby," he said in that rumbling baritone of his. "You're a Taurus, aren't you?" I was. "Mmm-hmm, I knewww it. My mother was a Taurus."

Here's the thing. When you're with someone like Barry White, a person who is so undeniably himself, you don't have to jockey to create atmosphere. He is the atmosphere.

He led me to the dining room, where a massive table had been laid out with a couple of iced buckets of champagne, crystal champagne flutes, bowls of caviar, and silver trays piled with toast points, and all the fixings: chopped egg, onion, capers. "What would you like, sweetheart?" he said. "It's beluga, baby." He took a seat at one end of the table, and I at the other, like royalty. A tiny white man, clad in a tux, stood stiff-backed against one of the walls. "The lady would like some champagne," White instructed, waving his hand, and the tiny white man hurried over to fill my glass. Our tuxedoed friend ladled caviar for us, too. "I'll take every-thanng," White said.

As the interview commenced, I asked White to list his three best qualities, and to my overwhelming joy, he answered in the third person. Who doesn't love a man who refers to himself in the third person?

After tenting his fingers and thinking for a long moment, he laid down his reply. "Barry White's three best qualities are his love for music, his love for people, and his love for himself," he said.

I was grateful for his booming voice, because I was a little hard of hearing after attending White's fabulous concert the prior evening, where I was surrounded by a group of Trinidadian middle-aged secretaries wearing the kind of bronze-sequined, shoulder-padded Cheetah-print separates that only a middle-aged Trinidadian secretary can get away with properly, and they would scream every time White would mop his glistening forehead with one of the many hankies that a disembodied arm would hand him from the side of the stage. The backdrop was an enormous bed headboard emblazoned with the initials B. W., and when he launched into "It's Ecstasy When You Lay Down Next to Me," he was accompanied by a coterie of lingerie-clad dancers who writhed in giant martini glasses. At the foot of each glass rested a small bottle of lotion, and the gals would periodically apply the lotion to their legs. That, friends, is a show.

Because of the tiny white man's vigilance with the bubbly, I was soon blind drunk, happily listening to Barry unspool his many philosophies. "Taurus, you've got to listen to what the universe tells you," he advised me. "So many people don't hear the truth when it's right there. It is right there." I nodded vigorously. He wound up the evening by counseling me on relationships. When you fall in love, he said, you lose contact with reality. Love as hard as you can, and as strong as you can, but never, ever fall in love. I thanked him for the advice. It was better than an audience with the Dalai Lama, and the Dalai Lama wouldn't have given me another big wet kiss when, hours later, I finally wobbled toward the elevator.

13.

One morning at *Rolling Stone,* Jann called me into his office. As his personal chef brought him a plate of fruit salad, he explained that Wenner Media and Disney had formed a synergistic deal, one of the components being that the Wenner-owned *Us* magazine would have a presence on the Disney-owned *Good Morning America.* Given my experience as a veejay, he thought I might be a good candidate to be the "*Us* correspondent."

My initial impulse was to leap up and run out of his office. I had, after five years, only recently stopped breaking out in hives on camera. I stared at him, fighting panic, as he outlined his plan. When Jann wanted something, he was the most charmingly persuasive person on earth. At the same time, of course, I was pleased that he thought of me.

I stalled by telling him I would think about it. The only TV I had done outside of MTV was two appearances on *Charlie Rose,* in which fellow veejays Carson Daly, Dave Holmes, and I were summoned to make predictions on Grammy winners. I could barely cope, so we first met at the Subway Inn, a Midtown dive near the studio that always had at least one patron snoozing with his head on the table and no lock on the door of the men's bathroom. The three of us knocked back a slew of drinks in quick succession. The other two were much calmer than I. That round oak table! The terrifying black backdrop! Those, those, those glasses of water! Set 'em up, Joe!

I went back to my desk and quietly made some calls.

"It's network," said Lou. "You'd be crazy to turn that down." He told me that I could probably ease out of my MTV2 contract. At that point, I was probably Viacom's oldest female veejay. It was only a matter of time.

"Do it," said my mother. "Jann has always supported you." Jann may have been a controversial figure in the publishing world, but he could do no wrong in the eyes of my mother, who had on two separate occasions sent him thank-you notes for being good to her daughter. In our house, Jann was nearly as important a figure as J. C. Penney. "And this is a chance to move into a more adult job. How long can you be on MTV?"

She was right. Maybe, as Elvis Costello once said, clown time was over.

Many meetings later, I agreed to try out. Jann wanted my *GMA* debut to be a report on "The Hunks of Summer," but at the last minute after calling around town in desperation, I was instead able to arrange a backstage piece on Madonna's flashy stage wardrobe for her *Drowned* world tour.

The amount of work that goes into one three-minute TV segment is astonishing. First, a producer and I spent hours at Madison Square Garden taping an interview with Madonna's wardrobe wrangler and exclaiming over her size-zero stage clothes. As we finally packed up to leave, the producer asked if we could borrow one of the black bras and panties that Madonna wore onstage to show Diane Sawyer, and we headed back to the show's Upper West Side office to write the script. After we wrote for about an hour, we waited for script approval from the higher-ups, after which we met with the script person who inputs Diane's intro into the TelePrompTer. Then it was time to record voice-overs in a little booth. The producer kept urging me to be more energetic, explaining that I had to sound over-the-top. With each try, I felt like I sounded more demented, but it's amazing how flat you can come off, otherwise, when the piece actually runs.

After an hour of this, the producer's work was just beginning. The process of editing our footage down, putting the piece together, and adding that clubby music that you hear in the background during the "action" part would keep him up until at least two in the morning. In the meantime, the executive producer had decided that I should appear with Diane the next morning to introduce the segment. A car would pick me up at five a.m. Oh, and wear a bright color.

I didn't sleep for a nanosecond. It was still dark when the car picked me up and glided into an eerily quiet Times Square. I stared numbly out the window at a gaggle of people who had already gathered in front of the glass-fronted *GMA* studio.

I was ushered into the greenroom, where a few staffers were pouring themselves coffee and hurriedly eating bagels. A production assistant rushed over. "There you are," he said. "I see you did your makeup."

"I forgot to ask if I was supposed to," I said, "so I just did it myself."

"Let's get you into makeup so they can put on a little more," he said. I was steered into a chair as the hair and makeup women descended, then back to the greenroom, where stood Dr. Phil, sipping a glass of water and talking to a colleague. He was due to go on next. The staffers had vanished and it was just the three of us, so I felt like I should say something to him. I wandered over.

"So are you a morning person, Dr. Phil?" I squeaked.

"Not usually," he drawled.

"Well, there's coffee over here!" I blurted, as though I had made it myself. I resisted the urge to slap my forehead. Dr. Phil smiled politely and resumed his conversation.

The production assistant reappeared. "You're going on at nine twenty," he said. Most of the hard news was covered in the first half hour, he explained. After eight, the softer stuff ran, when the viewer "wanted company."

I looked at the clock. Sixteen minutes to go. I drank more coffee. Oh, no. What was that fluttering in my gut? Was it some sort of stomach ailment? What happened if I had a problem right before I was supposed to go on? Would there be a shot of an empty chair where I was supposed to sit? I went to the ladies' room. No ailment yet, but I still had eight minutes. Dr.

Phil was called on. I watched his spot on the monitor in the greenroom. He seemed as comfortable as if he were in his den at home. When the show went to a commercial, my stomach lurched again.

"It's time," said the production assistant, leading me into a cavernous studio. Crew members scurried back and forth during the break, barking orders and adjusting cameras. He led me over to two armchairs. "You're in this one," he said. My stomach twittered alarmingly.

A soundman ran over and fastened a microphone on me. "Count to ten," he said briskly. I thought I was going to faint. Where was Diane?

"One, two, three . . . ," I recited. I felt like I was going under anesthesia.

A producer floated into view. "Diane is going to ask you two questions," she said. "You have about ten seconds to answer the first and about fifteen seconds to answer the second, can you do that?"

I nodded. Ten seconds, fifteen seconds. "What is she going to ask me?"

"Just general stuff," she said. "Don't worry, no curveballs." What was general stuff? What if I didn't know the answer?

"Can you give me an idea of what subj—," I said, but the producer had vanished.

"Twenty seconds!" someone shouted. I couldn't remember the name of Madonna's tour. The Girlie Show, maybe? No. What the hell was the name of her tour?

"Where's Madonna's bra?" another producer shouted. A p.a. raced over and whipped the bra and undies into my lap. They were shiny and black, made of some sort of techno material. I squelched the urge to examine them more closely.

A luminous, gorgeous Diane glided over and sat across from me as her mic was reattached.

"How are you?" she said, hurriedly straightening her shirt.

"A little nervous," I admitted.

She smiled warmly. "Oh, don't be." My heartbeat began to slow. As long as I kept my eye on her, I was okay. Keep your eye on the sparrow. Wasn't that a lyric from a song? It was. It was the theme from *Baretta,* which was sung by Sammy Davis Jr. That, I could remember, but when it came to Madonna, my mind was a sieve. The noise of the set receded and I started to drift away and float upward.

Up, up, up toward the ceiling. *God, Sammy was a true entertainer, and he never really got his due. He was so much more than the "Candy Man." He could sing, he could dance, he could act. He was a beautiful dresser. Sammy! One of the greats. Yes, one of the greats.*

Suddenly Diane was talking into the camera and our pretaped piece was rolling. I could barely pay attention to it because I kept wondering when I should hand her the bra and undies, and whether I should hand over the bra first, or both items together.

She was asking me a question about the clothes and somehow I responded, all the while thinking, *Wait, is this answer supposed to be ten seconds or fifteen?*

Then Diane remarked that I had brought something to show everyone and I found myself, absurdly, handing Madonna's underpants to Diane Sawyer, who made a self-deprecating joke, something about not being able to get it over one of her legs.

Then they tossed to a commercial, Diane thanked me and disappeared, my mic was removed, and three seconds later I was outside, blinking, on Forty-fourth Street, looking for the car to take me home.

A producer called me a few days later and said that I got the job, but first, a few issues needed to be addressed. I would have to go to a media coach to smooth out my rough edges. Also, my hair needed more "oomph."

"Oomph?" I said.

They sent me to a stylist who maintained the hairsprayed heads of many of ABC's on-air personalities. She snipped my long, wavy hair into the morning-television standard, a blunt-cut bob. I felt ten years older.

My clothes weren't right, either. At MTV, I had either worn my own clothing or outfits chosen by one of their stylists. At *GMA,* I had to provide my own wardrobe, one that was tailored, and, preferably, in jewel tones. My arsenal of black clothing wasn't going to cut it. The camera loves reds and blues and greens. I was dispatched to go buy a couple of suits with the show's wardrobe consultant, including a conservatively cut gray skirt and jacket. When I looked in the mirror wearing my new suit and senatorial hairdo, I was dismayed to find I didn't recognize myself.

Even with my new look, it was difficult to get pieces on the air. My tenure at *GMA* began right before September 11, so my ideas on celebrities and trends were, understandably, rarely considered (with the exception of Flag Fashion: Celebrities Show Their Patriotic Stripes). As the months rolled on and entertainment items slowly crept back into the programming, the edict was that spots had to be Water Cooler, which meant that people should be talking about them with their coworkers over the water cooler the next morning. Their infamous report of an eight-year-old child who still breastfed, for instance, was Water Cooler. And the ideas had to be newsy but not too edgy, as the average viewer was a woman in her midfifties.

I searched, frantically, for ideas. Jann wanted his "*Us* correspondent" on the air at least once a week. I could barely achieve once a month, despite writing long, frenzied lists of pitches on Celebrity Pets and The Return of Western Wear and Japanese Hair Straightening Techniques. I was told that the executive producer liked segments on items such as the season's must-have bag. I called stylists. I called stores. No must-have bag? Okay, how about a must-have shoe? Earrings—were there any must-have earrings?

Some ideas squeaked through, to my immense relief. One featured celebrity personal trainers who divulged the workout secrets of their stars. One interview took place at the Midtown studio of fitness guru Radu, who had worked with Cindy Crawford. When I shook his hand, his eyes flicked sympathetically over my physique. "I could help you," he said gravely. Another piece was on the emergence of butt cleavage due to low-rise jeans. We filmed it in the jeans department of Bloomingdale's. "It's not up top!" I shrilled as I walked toward the camera past a bewildered shopper. Then I looked wryly at her bottom. "It's down below!" Cue fast-paced "club" music!

Sometimes I would get a last-minute call to jump on a plane, once to Frankfurt, Germany, to interview Jennifer Lopez for her USO tour of Rammstein Air Force Base. Even though our chat was to take place in front of hundreds of keyed-up troops, I wasn't nervous in the least. Jennifer, resplendent in a white fur Chanel coat, was so dazzling that the troops were hypnotized. I could have been wearing a tutu and a scuba mask and no one would have noticed.

After my interview with Jennifer, I went back to the hotel and staved off homesickness by checking my e-mail. Two important missives were waiting from Lou. The first was a link to various recipes containing Jif peanut butter, including Jif apple pie and Jif soup, which basically involved adding broth and a few vegetables to a cup of peanut butter. *I would eat all of these,* he noted. The second was a passionate recommendation for an upcoming movie on Lifetime called *Sins of the Mind.*

> This is one of my favorites. It stars Missy Cryder, this actress who was once engaged to James Woods, who was much older than she was. Anyway, it's supposedly based on a true story and it's about this very responsible girl who gets into a car accident and hits her head and becomes a nympho. Her condition starts right in the hospital, when one of the doctors or maybe it was an aide, I can't remember, ends up giving her a beef injection, because that's how horny she is. And then she fucks her older, married neighbor one night when he's walking his dog in the woods, and then, the bartender at her sister's wedding. Her mother, who is played by Jill Clayburgh, eventually throws her out, so she goes to stay with a friend of the family, I can't remember his name but he used to be the head cop in *CHiPs.* She has sex with him, too. Her father, played by Mike Farrell from *M.A.S.H.,* is the only one who stands by her, and she doesn't even fuck him! At the end she has to go to a support group for nymphos. It was very moving. A must-see.

I dutifully entered *Sins of the Mind* in my date book. Oh, good: five e-mails from Julie. Every time I traveled, I did two things: told her goodbye, forever, because obviously the plane was going to crash, and reminded her that in the unlikely event that all went well, please send e-mails to cheer me up—the more mundane, the better. She obliged:

> For the last week I've been obsessing about the fact that I need a new sponge mop head, but for some reason, whenever I'm out I never feel like getting it. You know how that is—you don't want to go to the hardware store when you're out. Yet it's plaguing me. Sponge mop head sponge mop head sponge mop head. Only you would appreciate this.

Of course I did. One of the many behaviors we shared was a manic tendency to obsess about a task at hand until it was completed. If I had to go to the post office to pick up a package, I would remind myself dozens of times an hour that I must get to the post office. Post office post office post office post office. I was barely able to function until I hied down to the goddamn post office to pick up the mattress pad I had ordered.

Sometimes I would call her and say, "Pay bills online pay bills online pay bills online."

"What do I have for dinner what do I have for dinner," she would respond. "What do I have for dinner, tomorrow."

Germany was the only time that *GMA* sent me overseas. Mostly, I was dispatched closer to home, once to Los Angeles to chat with Robert Downey Jr. He had recently completed rehab and was participating in a charity concert organized by his ex-girlfriend. Ostensibly, we were covering the concert, but of course what *GMA* really wanted was Downey's comments on being in rehab. This is common practice among entertainment shows. You spend half an hour listening to the star talk about how they are doing it For the Children or For the Animals. Then you scrap that footage and use the throwaway comment tossed off at the very end about their new relationship or latest brush with the law.

Two chairs were set up at the huge house where the concert was taking place. Downey's publicist, whom I knew and liked, ambled over and firmly reminded me that Robert wasn't going to answer any personal questions, that this little setup was only about the charity event. In the meantime, my producers told me to lead off with a question about rehab, just in case he got upset and left. Various people filed in—I supposed they were event workers, or maybe pals of Downey's—and took their positions around the chairs, folding their arms.

When Robert showed up, he still looked a little fragile. He came right over to me and shook my hand. He was warm and appealingly goofy, joking around with me while the cameras were being set up. I felt like a crumb. I kept wondering how I could phrase the question that wouldn't make me look like a manipulative jerk. The producer finally told me to say, "Before I begin asking you about this worthy event, we all want to know"—assume Concern Face here and lean forward—"*how are you doing?*"

I took a deep breath and, looking everywhere but in his publicist's direction, did as I was told. Then I waited. Robert, without missing a beat, gave a thoughtful, measured answer. Thank you, Robert.

One-on-one interviews were preferable to working the red carpet at glitzy events. For me, a red carpet was a highway to hell. I just didn't have the required aggressiveness to corral stars and extract a quote from them.

For VH1's "Divas Las Vegas," all of the press gathered in one hot, stuffy room. TV crews are required to show up hours beforehand, and it's a long wait as you make cell phone calls and turn your face into a human crème brûlée, adding layer upon cracked layer of powder, so that your forehead doesn't glow on camera. I looked down the line of cameramen jostling to get the best position alongside eager, tense faces from the *Today* show, *Access Hollywood,* and *Entertainment Tonight,* their eyes trained to the door for the first sign of a publicist. Hours later, we were still waiting.

Here comes a publicist! We all surge forward. I grabbed my microphone and clutched my little list of all-purpose questions: Whom did you want to meet tonight? Who is the biggest diva of all time? The publicist scanned the logos on all the microphones, zeroing in on *ET.* Then Cher appeared, and I was knocked forward by a cameraman as everyone in the mob shouted and pushed. We were like hyenas spotting a limping oryx.

Unlike all of the other on-camera people, I tentatively stepped forward. "Get in there!" hollered my long-suffering producer.

"Cher!" I screamed desperately. "Over here! *Good Morning America*! I interviewed you for *Rolling Stone*!" Her publicist, thc famous Liz Rosenberg, spotted me and, to my delight, led Cher right over while everyone around me shot me poisonous looks.

The rest of the day was spent in the dressing rooms of Cyndi Lauper ("You can't have too many Q-Tips," she told me as she showed me her dressing table) and Shakira ("I love studded accessories"), after which we taped the entire concert, making notes on what were the most exciting or noteworthy bits (among them, Celine Dion's appalling cover of AC/DC's "You Shook Me All Night Long"). Then we hashed out the script ("Move over, Elvis! Last night was 'Divas Las Vegas' as VH1's divas show descended on Sin City!"), did voice-overs, and my producer spent the rest of the long evening in a satellite studio, cutting the footage. This, for a two-and-a-half-minute segment.

I was free to go after I received word that I didn't have to do any on-site commentary the next morning, the activity that I dreaded most of all. You go to a studio and an earpiece is fastened in your ear. Then, although Charlie and Diane are talking to you, you are unable see them. Instead, you are staring at a blue screen, trying not to look vague. When they pose a question—I always notice this now on news shows—there is usually a delay before the correspondent will hear. It is mere seconds, but when you're standing there, it seems like an hour. While you're waiting for the question to register, you are supposed to nod and pretend that you hear them so there isn't dead space where you're staring blankly, but inevitably you look vacant and slightly confused.

After a few more sporadic appearances on the show, my yearlong contract was nearing an end.

"I'm sure they'll renew," said Heather. "Lately you've really hit your stride. I loved that one you did on gift baskets to the stars."

As the last week came and went, I waited for the call. And waited. Maybe there was a heated meeting about it. Finally, one evening, a producer phoned me at home. "Listen," she said, "I'm sorry to tell you this but we're not going to renew your contract."

Somehow this was not what I had expected to hear.

"That's okay," I said automatically. "I appreciate your call." I hung up and sat down on the bed.

I was completely stung. I felt like I had finally learned how to do what it was that they wanted. I changed the way I looked, and talked, and acted. I fought the urge to call her back and ask what it was that they didn't like about me.

I didn't want to tell my parents just yet. They had taped every segment, sending out group e-mails to alert everyone when I was on. I hadn't eaten dinner, but I got right into my pajamas and slipped into bed.

Before I took the job at *GMA*, I had assumed that I could easily grasp being a correspondent. As it turned out, I could not. It was grueling work, and I gained a hearty respect for even the most small-town news correspondent. Still, I thought I had mastered it.

I lay in bed, trying not to cry. What now? Well, MTV was out—I was in my midthirties, so I might as well have been the Queen Mum—but I could certainly go back to *Rolling Stone.*

The next day, I vowed to be more upbeat. I hightailed it to the office to drum up assignments. Big bear hug from the mailroom guys! How are you, fellas? I headed down the hallway, waving at editors through their glass-walled offices. I spent so many late nights here, blasting my stereo unmolested as I finished up a story, gathering in someone's office for drinks before we left in a boisterous mob for a show. I chatted, I smiled. How's it going? Glad to hear it! Yep, I'm back! Well, thanks! Are you around later? I'll stop by!

As I neared the executive editor's office, the faces stopped being familiar. Who were all these people? Could things have changed so much in one year? His assistant, a young blonde in a midriff top, was new, too.

"Is he around?" I asked.

"He's at lunch," she said.

"Well, can you tell him I stopped in?"

"Sure," she said. "Who are you?"

Rolling Stone's fashion editor walked by. "She's a big-time writer here," she said, giving me a hug before she pressed on to her meeting. I affected a little "aw, shucks" shrug.

The girl looked at me blankly. A few interns, most of them college students, paused from their filing duties to stare at me with mild interest. Then they returned to their work. "Who are you listening to?" I asked, pointing to her stereo. "It sounds like My Bloody Valentine."

The interns exchanged glances. I used to flick that look at my fellow editorial assistants whenever some older writer would flap on about the time he saw Deep Purple at the Fillmore.

The editor had yet to appear, so I continued to linger, launching into a story about one of my *GMA* exploits, a report about the return of "big hair" in which we went to a New Jersey hair salon before the prom. As I nattered away, I noticed that the assistants around me rested their hands lightly on their computer keys, politely waiting for me to finish so they could resume typing in order to finish up their work before they haggled with one another about where to go that night. Just as I used to do.

I had become what I dreaded: an aging hipster, attempting to have a "rap session" with the kids.

"I should go," I said abruptly.

"I'll let him know you were here," she said brightly, turning back to her computer.

I walked quickly out. I heard someone call my name from one of the offices, but I kept going.

Drugs: Deciphering Your Metal Star's Mixed Message

Curious about drugs? Feeling distressed that the answer to "Have you ever been experienced?" is a resounding no? Resist the urge to talk to a musician about it. Even if it has been years since they touched anything illicit, even as they declare that drugs were responsible for their utter ruination, most of the stories that they will tell you from the old days will make drugs sound like lots of fun.

Further, some of the most enthusiastic ex–drug abusers can look surprisingly hale, which can be confusing. Ozzy Osbourne admitted as much during our chat.

"My life has been documented from fucking Day One, so everyone knows I used to take this drug, take that drug," Ozzy said. "I'm not a beacon of advice. When I try, the kids say to me, 'How old were you when you took your first hit of acid?' And I say, 'Yes, but look at me now.' And they say, 'Yeah, you're a successful man with a TV show.' "

14.

The more it was apparent that my days as a Rock Chick were waning, the harder I clung to the crumbling identity I had built up so carefully. I was going to plunge into the rock life while I could, and do it with abandon. It was time to rail against my square-shaped past. The most dangerous thing I had ever done was neglect to wear my seat belt on the Jersey Turnpike. If I hadn't met that girl from *Rolling Stone* at a party, I would be living in the Jersey suburbs right now, commuting an hour to an office park after I dropped my kids off at the Little Sprouts Day Care Center. Because of some twisted act of kismet, I was able to live every outlandish daydream I ever had in junior-year chemistry class. Wouldn't it be great if I could have the Ramones dedicate a song to me onstage? Or—I know—if I met Ray Charles and he offered to sing a song for me, so I pick "You Don't Know Me" and nearly weep at his heartbreaking a cappella version, just for me, in an elevator in Radio City Music Hall? Done, and done. When I lived in New Jersey, I had no idea what I was missing. Well, now I knew, and it was just too soon for me to return to the suburbs.

I started going out every night of the week. One night, at a party, I met Trevor, a dark-eyed, tattooed record producer with a thrillingly deep voice. I didn't tell my family about him—they wouldn't understand—but Julie might. After I had gone out with him for six weeks, I debuted him when the three of us met for lunch.

There was a frisson of tension in the air to begin with because I had been seeing less of Julie. Usually she and I talked on the phone a few times a day, but for the past few weeks, I had been dazzled by Trevor's nocturnal lifestyle. I dropped everything as I tripped from rock show to party to midnight dinner. Julie liked to be in bed by nine thirty. Suddenly it just never seemed to be the right time to call her back—either I was hungover, or running late for work.

After about five minutes, it was clear that lunch was going to be a bust. I knew that the two of them would not have much in common, but our little meeting was particularly awkward. Julie, normally able to chat with a tree stump, was unable to engage him. Trevor wasn't one for small talk, assumedly from logging long hours in a studio. I, who emptied out the contents of my metaphorical purse right away, was captivated by his mystery. It had been a challenge to pull information out of him—my ultimate interview. I covered it with nervous chatter and marshaled things along as quickly as possible.

"She's nice," he said afterward. "Not really my type, but I can see why you're friends." That was the extent of his editorializing on the subject. I admired his cool reserve. I wished that I, too, could sum up a situation with a few well-chosen words.

"He's okay," said Julie carefully.

"He's hard to get to know, but there's a lot to him," I said. "He's a big reader, for one thing. He was a medieval studies major."

"Well, maybe we'll go out again," she said.

My parents' encounter with Trevor was similarly dispiriting. When I couldn't put off our meeting any longer, we joined them for drinks (the first of many that evening, but they didn't know that) at a bar near the Met museum, where they had spent the day. Trevor was late (strike one), mumbled (strike two), and gave terse answers to questions about his parents, whom he usually saw biannually.

My father weighed in the next day. "In our opinion, he is younger emotionally than you are," he said.

"He's shy," I said. "You just dismiss everyone out of hand."

"He's stuck in being a twenty-something kid, even though he's in his thirties. I mean, you talk to him and he really doesn't have much to say about his future. And we think it's important to have somebody at least as smart as you are."

"He was a medieval studies major," I said hotly.

"I understand," he said. "It's just that when you look at couples when there's a big disparity in intellect and curiosity and things like that, you become bored with each other. That's the way I feel. And he never talked about his family unless we asked him about it. And you know we value family."

Oh, I knew, all right. I also knew this: The tyranny of being in a close family is that its closeness is jealously guarded. Woe to the outsider who upsets the equilibrium. "You know what, Dad?" I said. "Not everybody grew up the way that we did. I'll talk to you later." I banged down the phone.

Naturally, I saw even more of Trevor. We quickly fell into a routine: He'd finish up a recording session and then we would spend hours in dark, sticky-floored bars (whiskey for him, vodka for me), then stumble home when it was nearly light outside. I could talk shop with him without sounding like a name-dropping fool, and I loved that his reserve faded with each drink until he was slurringly effusive. In the mornings, I'd lurch awake, fighting dry heaves, and try to reconstruct the night's events. Where had we gone? I knew we'd argued for a while on the street, but then if I remembered correctly, we were making out.

Then I'd throw my grungy hair in a ponytail and roll into work at eleven, my hands trembling at the keyboard. Being with Trevor unlocked the tight control I usually had over my life. For the first time, I had the guts to actually live a rock-and-roll lifestyle instead of faking it. I didn't want to do anything except have Trevor whisper in my ear as we huddled in a shadowy bar. Drinks, shots, more drinks. Weeks dropped away, and then months. My skin paled until it was nearly blue.

My parents called occasionally, but it was easy to give them the slip because they were preoccupied with Dinah's new baby, their first grandchild, a girl named Claire. I started to chafe at their relentless wholesomeness. My father, the former Eagle Scout, whose life was so organized that he had Ziploc bags in five different sizes. When he discovered a dead bat during a routine inspection of the yard and wanted to take it in to the Department of Animal Control for "tests," he had the perfect bat-sized Baggie at the ready. My mother, the former head cheerleader, who was in her twenties during the sixties but had never so much as taken a puff of "grass." ("We were busy raising you kids, and you know what? I like to be in control. I'm happy with my wine.")

When my parents couldn't reach me by phone, a trickle of clippings arrived in the mail. The further I drifted, the more articles I received, usually with a note jotted across the top saying "Just in case" or "Worth a look." My father's clippings involved somber, practical matters ("It May Be Time to Roll Over Those 401(k)s, Planner Says," "Comparison Shopping for In-Network HMO Providers"), or, alternately, the many ways in which you can die in your own home ("The Perils of Throw Rugs," "Carbon Monoxide: Is Your House Harboring a Deadly Killer?"). Next to an article about a Roth IRA, he had cheerily scribbled, "I know you don't want to die old and broke! Love, Dad."

His assault continued on the computer with a steady supply of e-mails forwarded from his JCPenney crony Vern Leister, with titles like "Ever Wonder (. . . why is it that doctors call what they do 'practice?' . . . why women can't put mascara on with their mouths closed?)" or "Only in America (. . . do we buy hot dogs in packages of ten and buns in packages of eight!)." SO TRUE, my father typed.

My mother stuck to clippings, but hers fell squarely in the lifestyle or human-interest category ("Scrapbooking: More Than Just Memories," "102-Year-Old Man Says 'Don't Sweat the Small Stuff,' " "Sensible Sandals

Can Combat Arch Enemies"). Usually she scrawled the word *Funny!* on top. I put them aside. My life wasn't that easily managed.

I discovered that in New York, whole weeks can glide smoothly by without anyone knowing exactly what you're doing. To preserve the illusion of normalcy, I would check in with friends for a quick coffee, but all of my nights went unobserved. Dinah was a new mother. Heather had recently married Rob, and both were working in restaurants in the city. I had never even met my next-door neighbors. No one was watching. What drew me to Trevor was that he didn't seem to need anybody. He hopped from studio to studio, hanging out with whatever musicians were in front of him at the time. He rarely spoke to his family, while I had always fielded a dozen phone calls a day from mine on urgent matters such as Did You Send Your Cousin a Card for Graduation or Should Dinah Paint the Deck? I wanted a life like Trevor's. People floated into his life and then floated out. He did whatever he wanted.

I had a few conversations with Julie, but they were hopelessly strained. Even though we had known each other for a decade, we didn't have a confrontational friendship. In the past, if she hadn't heard from me, she would never get angry, just concerned. Our relationship was based on complete support, so my addiction to being with Trevor was something that neither of us knew how to handle.

I attempted a shopping date. We went to the lunchroom at Bergdorf Goodman, our usual haunt for lobster salad sandwiches with the crusts cut off, but I was restless. We cruised the racks, always easy territory.

"What do you think of these?" I said, holding up a pair of black leather pants.

"Well, you can get away with them," she said. "You're really thin. This is probably the thinnest you've ever been."

"I just don't feel like eating lately," I said absently.

One night, Trevor had to meet some musician buddies at Maxwell's, the rock club in Hoboken. Dinah's apartment was ten blocks away, but I didn't want to call her. I just couldn't picture her interacting with my scruffy, chain-smoking group, with her baby pictures and flowered skirt and chirpy questions.

Outside the club, a bass player whose name I gathered was Moz showed Trevor a sizable packet of coke that he had hidden in his leather jacket. "It's gonna be a late night, dude," he said, cackling.

"Sweet," said Trevor.

I loved the furtive looks that we got as we all slunk through the door in a dark-jacketed pack. We started the night with Jack-and-Cokes. I hadn't talked much because I was intimidated by Trevor's friends and their offhand chat about guitars and gigs, and when Moz followed me into the bathroom and tapped me out a couple of frighteningly fat lines, I decided to go ahead and do it. Maybe it would loosen my tongue. I had never liked drugs, but I was sick of being the good girl. Where did it get me? So pitifully eager to be liked, scrambling to please everyone, feverishly flapping my top hat up and down for my editors, my parents, celebrities.

I grabbed the straw and vacuumed up everything I saw.

"Whoa," said Moz. "That's way too much."

I kept going.

My heart began to hammer sickeningly. *Steady, now,* I told myself, but as I looked down and saw my chest pounding in triple time, visibly pumping through my shirt, my fright increased. "I need a little air," I mumbled as I pushed my way out of the bathroom. Trevor and another friend were already queued up to help themselves. I felt light-headed. What if I passed out? My cheeks burned. My body temperature abruptly spiked, and I was scalding hot. I pushed up my sleeves and saw that my arms were bright red and I knew without seeing myself that I had flushed a deep scarlet. People were staring.

I made for the door. I had to cool down. Fortunately, it was January, and the temperature hovered in the teens. I ducked onto a corner street and sat on the curb, trying to slow my breathing, but I was hyperventilating. Why wouldn't it stop?

I fumbled for my cell phone and called Dinah. I got the machine. I pictured her quiet apartment, with its black-and-white checked floor and hyacinths blooming in a pot, and her two old gray cats snoozing on the bed.

"Dinah," I said. "Oh please help me." I started to cry. "I need you. Where are you? I'm at Maxwell's."

I snapped the phone shut. My face was slick with sweat, and my heart beat ferociously. Where was Trevor? He and his friends had forgotten me utterly, an egocentric by-product of the drug. Who could I call? Who would help? I couldn't phone Julie. Blindly, I thought of calling the police. A dark patch flowered on my shirt. My nose was bleeding.

A cab lurched to a stop in front of Maxwell's. It was Patrick.

"Stay here for a second," he said to the driver. "I've just got to find someone."

"I'm over here," I called from the side street.

I was sitting with my arms wrapped around my knees. Patrick squinted into my face.

"How ya doing?" he said. "Not so good, huh?"

I tried to make a joke to alleviate how scared I was, but I couldn't. I cried harder as blood burbled out of my nose. "I did a lot of blow," I said between gasps. "I don't know what to do."

"I've got a taxi here," he said. "Why don't you get in the cab and we'll go to your place." He picked me up awkwardly. We weren't entirely comfortable with each other. I realized that I had never actually been anywhere with him alone. In my family, every activity was done in pairs. If I had to run to the supermarket at a family gathering, it would never occur to me to ask Patrick to accompany me. I would bring one of my sisters. You don't break up the unit.

I could tell by his gruffness that he was as frightened as I was.

"Talk to me," I said, crying.

"We're almost at the Holland Tunnel," he said. "I don't have any tissues, I'm sorry. I should carry a hankie, like your dad does."

"Keep talking," I said. "It calms me down." I tried pinching my nose but the blood kept flowing, mixing with my tears. "I can't believe I did this," I said miserably.

"It's okay," he said, petting my arm awkwardly. "It'll get out of your system soon. Uh, let's see." He searched for something calming to say. "Dinah is with Claire at your parents' house, getting a little free babysitting."

As the taxi turned on my block, my heart began to beat more normally. The cabbie, well used to seeing this sort of behavior, discreetly provided some tissues for me to wipe my nose. We pulled up to my apartment and Patrick helped me out of the cab.

"You okay?" he said. "You look a little better. You want me to come in?"

I shook my head. "No, that's okay," I said.

He persisted. "You want me to make you some tea or something? I don't know anything about being on coke but maybe you need to get something in your stomach. Maybe I should make you a sandwich."

"That's okay," I said, but I paused. Suddenly I didn't want to be stoic. "Actually, could you please come upstairs?"

Together, we slowly walked up the stairway, then he deposited me gently on my bed. I cleaned all of the blood off of my face and shakily climbed into my pajamas while he bustled in the kitchen. "There's not much in your fridge," he called. "Don't you ever cook?" I went out every night. I never made anything anymore except cereal for breakfast.

A pause. "Here's some bread. Okay, I can at least make you a grilled cheese."

I sat at my tiny kitchen table. "I'm really embarrassed," I said, sniffling. "I don't know what I was thinking. I had been drinking so much that I guess it warped my judgment."

He found a frying pan and slapped the sandwich onto it. "I understand why you're embarrassed," he said over his shoulder. "But let me tell you something, I could tell plenty of stories that would make you a lot less embarrassed." He laughed. "If it makes you feel any better, I've done things that make what you did tonight feel like nothing." Patrick, a former enthusiastic drinker, had been sober for years. At family gatherings, when my folks broke out the wine, he always had a Virgin Mary. "I mean, I've been sober for a while now and I'm very proud of that, but I struggle with it," he said.

He flipped my sandwich over, then ran to get me a blanket. "Put this around you," he said. "I remember one time, it was around the holiday season, I was working as a sous-chef at the Tudor Hotel on Forty-first Street. The hotel was having their employee Christmas party, but I was supposed to go home because Dinah's friends Sarah and Mark were celebrating their anniversary. Because I was a chef they had been bugging me to make them a meal, so I had promised to make them a gourmet dinner in Hoboken." He flipped the sandwich onto a plate and pushed it toward me. "So I was going to leave the party because I knew it was a potential disaster," he said, taking a seat across from me. "But my boss, the chef, talked me into staying for 'just one drink.' And as any substance abuser will tell you, 'just one drink' or 'just one sniff' are the famous last words."

He shook his head with a bemused smile. "So the bartender was sneaking me straight vodka at this holiday party and I just got hammered. I probably drank easily a pint of vodka—sixteen, seventeen shots."

I stared at him. "Oh yeah," he said. "I was getting so drunk I was putting out cigarette butts right on the nice new carpet. At which point—I don't remember this, but I was told later that the general manager whispered in my boss's ear, 'I don't care how much it costs, get him in a cab and get him home.' I don't even know how I got home, don't remember coming home. Don't remember anything."

He folded his arms. "What I had done was stagger in the front door while Dinah and her two friends were sitting in the living room. I was three hours late and not only was I supposed to cook but I was supposed to buy the food. I didn't even say anything to them because I didn't realize they were there. I stripped down butt naked, in front of them. Because it was French doors. Okay? And French doors have fifty windows. I opened the French door, walked by them to the kitchen, stuck my head under the kitchen faucet and got a drink, walked naked back to the dining room, gave Dinah a kiss, and said, 'I'm beat, I'm going to bed.' And then I went to sleep." He shook his head again. "At which point Mark turned to Sarah and Dinah and said, 'We need to get him to the gym.' "

He laughed loudly. "What a dick!" he said. "I think that upset me more than anything else."

Then he looked at my grayish face and his smile faded a little. "Uh-oh," he said. "Are you going to get sick? Let me help you." He struggled to pull me up so I could run to the bathroom. "It's okay," he said as I grabbed onto his arm. "Not a big deal. No, no, don't cry. There we go."

The next day, Dinah phoned me at work. "I heard what happened," she said carefully. "How are you feeling?" Oddly enough, I felt fine. My vows from the night before—*I will never, ever do this again*—were already beginning to fade, especially after a contrite Trevor showed up at my apartment with bagels and coffee, telling me sorrowfully that he had been walking up and down the streets of Hoboken, looking for me. In the mild light of morning, the whole episode had transformed into our first drama, one that we weathered together.

"I'm okay, thanks to your husband," I told Dinah.

"Listen," she said. "Can you just please be careful? I'm worried about you."

"Don't you worry," I said firmly.

"I wish I had been there," she said.

"Really. Honestly. Don't worry."

"Okay," she said. "Well, I won't lecture." Dinah didn't push. I still had some residual status as the eldest sibling. "Listen," she said. "I wanted to talk to you about Thanksgiving." Our family celebrated every single holiday together, including all birthdays. Each one had its own set of traditions. The night before Thanksgiving, for instance, we always went to an old-fashioned Chinese place near my parents' house that was shaped like a giant pagoda and had a Plexiglas bar with goldfish swimming inside. We ate heaping plates of dumplings with peanut sauce and read our fortune cookies aloud.

Christmas Day involved some sort of family craft—one year we made birdhouses, another time we did gingerbread houses that my folks donated to the local senior center.

This year, the thought of Thanksgiving exhausted me. I just wanted to dislodge my family's tentacles. Dinah always called a month in advance to "organize" everything, even though it was the same drill every year. She loved to plan. "So are you coming home Tuesday or Wednesday?" she asked. "Because if you're coming up Wednesday, then we'll come up Wednesday afternoon, but I need to know now because we need to know what room to put the baby in, and if we go to the Chinese restaurant I want to get a table around six, or even five thirty, because Claire has to go to bed early."

I reached into my drawer for my econosized bottle of aspirin.

"Fine, that's fine," I said. "But I really wanted to bring Trevor home. I have to talk to the folks. I don't think they're too keen on him. Let me call you back."

I dialed my mom. "Hi, stranger," she said.

"Hi, Mom." I girded myself. "Listen, can I bring Trevor home for Thanksgiving?"

There was a short silence. "Doesn't he have his own family?"

I gulped down three aspirin. "His parents are going to the Bahamas and he doesn't have anywhere to go."

She sighed. "You know, this is really a time for family. You really haven't been dating him that long." I felt the first prickle of irritation. Why did I keep bringing these guys home? Why did I bother?

"I thought the idea of Thanksgiving was that you were supposed to be inclusive," I said. I could feel the peevishness creep into my voice.

"I'd rather you didn't," she said.

I closed my eyes and rubbed my forehead. "Mom, I know you keep hoping I'm going to bring home a Wall Street guy, but it's not going to happen."

She was quiet. "That's not what I'm hoping. I just feel like what you want isn't necessarily what you need. There is just a kind of disconnect between what you see yourself as and what you actually are. You know?"

I sighed. "You know what, Ma? Maybe I'll stay home this year. I'll actually be on deadline the day before Thanksgiving, anyway." This was a major family transgression. I had never missed a single holiday, even when I was in college.

Another silence. "Well, do what you have to do," she said curtly.

I called Heather, hoping that my cubicle mates in the office couldn't hear me. "What are you doing?" I asked.

"I'm trying to get out of the house, but Rob won't take his nose out of his new cookbooks." Rob loved his work as a chef, so much so that he collected cookbooks by the hundreds. "He just got three of them on all these hot new Austrian chefs," Heather said. "He thinks if he gets them online, I won't know. And then he does what I do when I go clothes shopping—he subtracts fifty bucks from the total when I ask how much he paid. Or he says that the books were on sale when they really weren't. Or that he had a store credit when it actually ran out two years ago."

"Listen," I said. "You have to work at the restaurant for Thanksgiving, right?"

"We both do," she said. "You know how I love holidays. That's the one bad thing about this stinkin' business, but what can you do. But the night before, Rob and I are going to have a nice dinner at home." Rob always made jokes about Heather's enthusiasms, but I could tell that he secretly enjoyed all of her lavish rituals.

"Well, I'm not going home," I announced. "They don't want Trevor to come, so I'm not going to, either."

"Well, don't worry, it could be great," she said. "Go to the gourmet store and get a nice little Cornish game hen and some mashed sweet potatoes and a pecan pie for two. Get dressed up, light candles, get a bottle of champagne, put on some Nat King Cole, and you can have a romantic dinner. Oh! I know! Go to a card store and get one of those paper turkeys that you open up in the middle of the table. Then you can make a nice Thanksgiving for him since his parents aren't around."

When I hung up, I felt a little better. I always did exactly what Heather prescribed. The day before Thanksgiving, I bustled around the city while Trevor worked a long shift at the studio, picking up a pie at the Little Pie Company and racing to three different card stores before locating a paper turkey. I never minded the long lines during holidays because everyone was in a good mood, for once. I even got into an impromptu conversation with an ancient man in a bow tie with three hairs combed heroically in an elaborate swirl on his head. He was as leathery and desiccated as King Tut's mummy, but he was a vigorous advocate of corn bread stuffing over plain white bread stuffing.

I came home to a message on my machine from Julie. I hadn't talked to her since she had returned from a vacation in Italy. "I told my mom that when I was on vacation I had an epiphany," she said. "I saw a woman and thought, 'She should be wearing a one-piece.' Then I thought, 'I'm not going to think that way, and when I have a daughter, I'm not going to pass that kind of thinking on to her.' And you know what my mother said? 'I was in Florida this winter and there were all these fat people walking around in bikinis.' That's what she got out of the conversation. That was her response. As usual, she crystallized my thoughts perfectly. Anyway, you might be in New Jersey but call me if you're around."

I must call her today. I will, after I've made dinner. Later. Definitely.

"I'll do everything," I told Trevor when he arose at eleven. "You just relax."

He smiled bemusedly and picked up a newspaper as I set the table. "I hadn't planned on doing anything," he said.

I switched on the TV so that I could see the parade.

Trevor put down his paper. "What are you doing?" he asked.

I shrugged. "At home, we always keep the parade on," I said. "I'll turn the volume down."

"Can you just mute it, please? Thank you," he said.

I grabbed the remote. "Sure," I said. I put the game hen and the sweet potatoes in the oven. Should I make yeast rolls? It was just the two of us.

"Smitty called," said Trevor. "He wants us to meet up with him after rehearsal."

I stopped flipping the pages of a cookbook. "What? Tonight? I was thinking we could watch old movies or something."

He laughed. "Jesus, it's just Thanksgiving. Come on, let's go have fun." It occurred to me that Trevor was thc one who drove the bus in our relationship. When did that happen?

"Okay," I said.

After an hour, the food was ready. It's not like a game hen takes a long time to cook.

"What's this?" Trevor said, sitting down at the table and pointing at the paper turkey. "Why would you spend money on that?"

I spooned up some sweet potatoes. "Heather thought it would be festive," I said. Why did I sell her out like that?

I had asked him to dress up. We were both wearing head-to-toe black, as if we were attending a funeral for the game hen. As I picked up my fork, Trevor jumped up. "Almost forgot," he muttered. He searched his knapsack and brought out a joint. "Want some?" he said with a grin, lighting it up.

"No, thanks."

"You know me," he said. "I'm not a big eater. This makes food taste better." I flashed onto my parents' house. My mother was probably fussing over the elaborate autumn display on their Thanksgiving table. She rotated each one, carefully noting it in the calendar ("Thanksgiving 2000: Indian corn, white ceramic pumpkins, mums in center. Thanksgiving 2001: orange and yellow squash, maple leaves, gold candles").

The phone rang. It had to be someone in my family. I let the machine pick it up. I was obviously having so much fun that I couldn't possibly answer the phone.

It was Dinah. "Where's the recipe for your yeast rolls?" she asked. "I thought it was in this old *Better Homes and Gardens* cookbook. There's so much flour on it, it's disgusting, it's like papier-mâché." I could hear her flip the pages. "Hey, where are you? Is anything even open? I know. You're probably at the parade. Well, I missed you. We didn't do a craft this time because there weren't enough people. You're in for Christmas, right? Please tell me you're in. I'll probably be up until ten. Okay, maybe nine. Call me."

I tried to eat a few sweet potatoes but the lump in my throat prevented me from swallowing. What was I thinking? Trevor, who rolled his eyes whenever I talked about my family, would have loathed Thanksgiving at my folks' house. He would have been twitching and mouthing *Let's go* at just about the point when all of us went around the table and announced what we were most thankful for. I looked down at my plate, trying to maintain control. I had thrown up in front of him after one whiskey-soaked night, but I had never cried.

It hit me: Being hip was a full time job, and I was only a part-timer. The rest of the time, I wanted to put red hots on a gingerbread house with Heather and Dinah at my folks' dining room table. I was a geek. When I first met Trevor, he told me teasingly, "I'll bet you're a cat person." I denied it, but the fact stood that if I happened to be at a newsstand, I surreptitiously peeped at *Cat Fancy* magazine as though it were a dog-eared copy of *Juggs*. I couldn't hide forever that I liked county fairs, particularly the goat booth at the 4-H tent, or that I once spent a week with my grandmother at her house in the giant retirement community of Sun City, Arizona, and it was one of the most carefree times of my life.

Every morning "Ma" (she didn't like the term *Grandma,* which made her feel old) would pad into my room in her slippers with a glass of orange juice for me. Then she would pilot her mammoth, cream-colored Buick to thrift shops, where she was always on the hunt for elasticized polyester pants in pastel colors. After pawing through the racks, we'd indulge in a buffet lunch at her favorite Mexican place or, sometimes, Chuck E. Cheese's, because she liked to "watch all the young people."

Then, after a trip to the mall, we'd head back to her house, where she would show me various favorites from her extensive doll collection, or maybe my great-aunt Lucile would come over for "a nice visit," or, if I was feeling particularly wild, I'd put on Ma's flowered bathing cap and stroll down to the community pool for a little aqua-robics. Ma would make dinner promptly at six, after which we'd retire to the living room and thumb through copies of *Ladies' Home Journal* while we watched *Murder, She Wrote.* After every meal, we helped ourselves to dessert (candy bars, she declared, were good "for strength").

I was a geek, with a healthy dose of Old Lady. 'Twas ever thus, and I couldn't keep pretending forever. Guess what, Trevor: Soon enough you'll discover that lurking behind my Wire albums is Rick Springfield's *Working Class Dog.* Maybe we'll put that on the turntable at your next party.

Trevor picked up the game hen. "Don't eat me!" he squeaked, making it wiggle around as if it were running.

How to Write a Sex Column Despite a Distinct Lack of Field Research

Your first order of business is to compile a database of different ways to say "have sex." Some useful terms are boinking, making a deposit, doing the humpty dance, getting parallel, raw doggin', having a bit of the old in and out, lancing, swapping gravy, oofing, and—my personal favorite—spelunking.

This will avoid redundancy. It is also helpful to accumulate inventive ways to describe a man's equipment, because you will be referring to it a great deal. Some good ones to try are schwantz, hose, anaconda, bratwurst, badajo, kickstand, flagpole, joystick, and—my personal favorite—cob.

Finally, if your sex life is about as exciting as a televised oil-painting show because you have recently broken up with your boyfriend, Trevor, ply your more adventurous friends for material.

Dear Dr. Sooth,

My friend swears he can tell a woman's nipple color by looking at the inside of her lower lip. Is this possible?

I stared bleakly at the computer screen. Ten minutes passed. Twenty. A crony of mine had recommended me for the job of *GQ*'s sex columnist, and despite my stunning lack of credentials, I took the gig, especially after I learned that I could hide under the pseudonym of Dr. Sooth, which was some sort of amalgam of "soothsayer" and Dr. Ruth.

Readers were to send in their questions, which were forwarded on to me. Then, being spectacularly unqualified to answer the question myself, I would phone up an expert. For some reason I thought that I would receive bags of letters, like in the scene at the end of *Miracle on 34th Street*, but usually there was only a trickle. I was puzzled until I met a fellow sex columnist at a party (New York, in its way, is a small town).

"Please," he said. "My column is aimed at both sexes, but ninety percent of my letters are from women." He laughed. "Most of the time, men assume that nothing is wrong and that they're doing just fine."

Worse, the few letters that I did receive had a distressing uniformity, usually along the lines of *How can I pleas* [sic] *a woman?*

Or, alternately:

I was wondering how I could make my dick bigger. Not that the ladies complain. Thank you. Peace out.

It never seemed to occur to the advice seekers to wonder why, if I knew the magical secret to enlarging a man's cob, I would be slogging away at a sex column.

Sometimes, however, a doozy would arrive from heaven, like the question about nipple color (I checked with my doctor, who torpedoed that theory by saying that there was no pigment inside the mouth). So much to learn about my fellow man! His colorful proclivities, his wonderfully exotic foibles!

A friend of mine claims that most men put on a condom before receiving a lap dance to prevent stains. Can you fill me in? ("That's a new one," snorted one of the many strippers I consulted on matters of adult entertainment.) *My johnson is covered with hair, almost up to the tip. Can I use laser hair removal on it?* No, you cannot, as no hair-removal emporium will go near genitalia due to a justifiable fear of lawsuits. Buy yourself an at-home waxing kit.

I have always wanted to try water sports. How can I get my wife to go along with it?

"Water sports, commonly known as 'peeing on each other,' " I wrote insanely, "are harmless, as long as they are consensual."

I met a girl through a personal ad who seemed perfect for me, but when I went to her house to pick her up, she had a huge macaw perched on her shoulder. That freaked me out a little bit, but I still thought she was cute, so I put it out of my mind. After we went out for dinner and drinks, I go back to her place, one thing leads to another, and we had sex on the couch. The problem is that that goddamn macaw was watching the whole time and it gave me the creeps. What do I do? Also, the bird sheds "macaw dust" all over everything.

Naturally, I speed-dialed the proprietor of Dick's Macaw World in Thomasville, Pennsylvania, the preeminent macaw expert in the Thomasville area. I had the good fortune of getting Dick right on the phone. Perhaps unsurprisingly, he angrily sided with the macaw. "When you've got a macaw, you've got a friend for life," he huffed. He recommended that the guy make nice with the bird, said that macaw dust was harmless, and added that macaws can live dozens of years. "Longer than most relationships," he said pointedly.

I sighed. "Dick," I said, "you're absolutely right."

15.

It was time that I started fixing my own affairs. I phoned Julie, dialing her number quickly so that I wouldn't lose my nerve. "Can I come to your house?" I blurted when she picked up the phone.

"Sure," she said. I could tell she was treading lightly. "I'm just going to walk the dog and then I'll be here."

I hurried up to her house, light-headed with nerves. We hadn't had a real conversation in half a year.

When she opened the door, her face was a slide show of different expressions: happiness, doubt, suspicion. Her new dog, Otto, capered around my feet, but she stood still. "Hi," she said.

"Oh, Jul," I said, and gave her a hug. Tears squeezed out of my eyes. "I'm so sorry. I've been such a jerk."

"Come on in and let's sit you down," she said. "You need a Kleenex?" I shook my head and took a seat.

We sat, facing each other. "So how are things?" she asked lightly. "How's work? How is Trevor?"

"It's over," I said, wiping my eyes. "I told him I didn't want to go out as much. All of those late nights just started to get old. He continued to go out, of course." I shrugged. "And then one time I realized that I hadn't seen him in a week. Then two weeks. He never did want to stay home, even though I was there. Maybe he wanted a drinking buddy more than he wanted me. Anyway, the phone calls kept getting farther apart until they stopped altogether." After Trevor was gone, I felt like I had recovered from the flu—I started getting up early again, and my skin lost its sallowness—but a fog had descended around me that just wouldn't lift.

I sighed. "I know no one liked him. Which, of course, made me rally to his side, rather than realize that perhaps it's worth paying attention if someone is reviled by everyone you know."

Julie smiled wanly. "You'll probably start dating him again in a month and you're going to hate my guts, but I just have to tell you that it seemed to me that he was only in it for what you could do for him." She sighed. "Remember your birthday party a few months ago? I had gone with my dog to get your cake at the Cupcake Café, which for some reason is right underneath the drop-in center for the homeless shelter near Penn Station. I didn't want to tie up Otto because those people would eat him alive. You know how difficult it is to pick up a gigantic cake and keep the icing flowers nice so they don't get smushed with a dog?"

She looked at me fiercely. "Trevor did nothing for the party and put no money in, and at one point you were thanking him and he had this shit-eating grin on his face, taking credit for the whole thing, and I hated him."

"Why didn't you tell me this?" I said.

She raised an eyebrow.

"You're right," I said. "It wouldn't have made a difference."

She went on. "I remember being in your apartment after the party was over and we were getting ready to go to a bar, and you and some girls from *Rolling Stone* were in the bathroom putting on makeup and I was on the couch," she said. "You were all wearing Daryl K low-rider pants and you were super skinny and I was hiding my extra ten pounds in an Eileen Fisher outfit that my mom had picked out, and you were all looking in the mirror and laughing and I didn't even know what was going on in there. I was completely out of it." She shrugged. "That was the beginning of the end for a while. I saw that you had turned into a rock and roller. It wasn't bad, but one of the things that we bonded on was that we were very cool people, and lots of people wanted to be friends with us, but we loved our geeky parts. You know? And you just seemed to turn on that."

I nodded. "I don't know how to make this up to you," I said.

She waved her hand. "Let's forget it," she said. "I just miss talking to you every day. There are so many things I wanted to tell you. I got a job, for one thing, at VH1. I'm going to be a writer at *Pop-Up Video*—you know that show? I start in a few weeks."

"What? Oh, Julie. That is so great." I fought back another gush of tears as she filled me in. Four hours later, I was still in the same spot on her couch. When I finally rose to leave, I gave her another hug.

"I'll call you later," I said.

As I rode the subway home, it occurred to me that I had no plans. After my protracted Lost Weekend with Trevor, I had begged off of plans so many times that I received few calls. My pride prevented me from trying to reconnect just yet. Maybe it was time to stay in for a while.

I got off at my subway stop and headed to a gourmet takeout place that was popular with all the single people in my neighborhood. I never cooked for myself, only for others. Left to my own devices, my dinners were random assemblages. To celebrate the relief I felt at seeing Julie, I decided to get a decent meal. Maybe I'd go completely crazy and get a few side dishes.

I joined one of two lines and furtively checked out my fellow singletons: four women in their thirties, and one frat boy type, still in his suit from working on a Saturday. I caught the eye of one of the women and we exchanged a look. *Single? Right. Me too. Let me guess, you're going home to eat that Savory Tamale Pie in front of the TV. And you'll do the laundry because on a Saturday night, the laundry room is empty. Oh, you have a cat, too? What's his name? Mr. Purrbox? That's cute.*

I shook my head to clear it. Why did I have to cast things in such a gloomy light? Maybe those people were all perfectly happy being unattached. I certainly had enough married people tell me that they envied my footloose life. Heather always accused me of attaching a dismal backstory to anyone I saw who happened to be alone. If I spied an older man having a solitary meal at a coffee shop, I inevitably thought, *Sad widower, directionless since his wife's lingering death from cancer two years ago, spends his allotted monthly splurge from his Social Security check at the roach-ridden diner before going home to his ammonia-and-fried-onion-scented apartment, where he has cold tea and wonders how to most efficiently end his life.*

"Maybe he's wealthy and likes going to coffee shops," Heather would point out. "Maybe he is finally free to eat cheeseburgers and watch sports all day since his nagging wife is gone."

I wandered out of the shop. What to do? I walked aimlessly along before ducking into a grocery store. Grocery stores always cheered me up, even the cramped Third World one in my neighborhood. I grabbed some cereal—my usual lunch and dinner—and a few pears and got in line.

The cashier scanned my food without looking at me. She stopped at the pears, frowned, and picked one up.

"What are these?" she asked.

"Bartlett pears."

She sighed. "Cliché, I need a price check on Bartlett pears," she hollered to a coworker.

I looked blankly at her.

She slammed down my pear. "Cliché!" she yelled. What was she talking about?

I glanced over at the cashier next to her. Her nametag said CLISHAY.

Fantastic.

I ran home to tell Heather.

"I have a new favorite name," I said, flopping on my bed with the phone.

"That's good, because I'm in the market for names," she said.

"What do you mean?"

She laughed. "Well," she said slowly. "I'm pregnant."

"Heather!" I said, jumping up. "I love it!"

"Yeah, yeah," she said. "Whatever." Heather downplayed any of her major life events. Like the Brits, she was discreet, had a caustic sense of humor, and ran from overt attention. "It sort of happened on our first try, believe it or not," she said.

I slipped into my pajamas, even though it wasn't yet six o'clock. Why not? "Are you excited?" I asked.

"Yes. Seriously, I really am. And Rob is especially excited. He's so ready to have his own family. He's already told the world. The dishwasher at his restaurant knows, and I haven't even let Mom know yet. He already bought a little photo book to carry around in his bag." She sighed. "The only downside is that I just feel sick all the time. The one thing that seems to help is pressing my face up against the cold file cabinets in the office at the restaurant when no one's around."

"God, my little sister is pregnant," I marveled. "All of a sudden I feel as old as Methuselah."

"Yes, but look what you do for a living. You travel all over and interview all these interesting people. You can do whatever you want."

I sighed. "I never thought I'd say this, but sometimes it gets a little old to do whatever you want, all the time," I said.

Silence. "I can see that," she said finally.

"One day I woke up and I was in my midthirties," I said. I told her about my cringe-inducing visit to *Rolling Stone.* "I have just been thinking a lot lately about what my plan is," I admitted. "I really have no idea. What's your plan?"

She thought for a minute. "I've always seen myself with kids and a family," she said. "I'm a homebody, I always have been. I guess if I had a plan, it would be to relive everything I grew up with—catching fireflies and

Flashlight Tag and playing dress-up and birthday parties. Going to get ice cream after swim meets. Divvying up my Halloween candy and trading away the boxes of raisins and the Tootsie Rolls that you get a hundred of. I don't know, I see myself as a grandparent with lots of family around. So I guess it depends on what you want."

"Boy, Heather," I said. "You sure know how to create a mood. Well, I'm going to think about it for a while. And in the meantime, no more bars."

"Good. You don't want to get haggard. Stay in, and read, and take baths, and write, and take long walks. Eat dinners out by yourself. Ask Lou to come over. He always makes you laugh."

He came to my apartment a few nights later, dressed in sweatpants for a night in. "I brought you a Whitman's sampler," he said, tossing it onto my kitchen table. "Everyone loves a Whitman's sampler. And it's no mystery—they have a diagram inside of what's in each chocolate. At a time like this, you need certainty. Nothing's going to let you down."

He held up a tape. "I also brought over a movie that will really resonate with you right now. It's called *Touched by Evil,* starring Paula Abdul in her first and only TV movie. She plays this single career woman, like you. She doesn't want to be in a relationship because she had a bad experience, but then she meets this dashing stranger, played by Adrian Pasdar, who in real life is married to the squat one in the Dixie Chicks. She opens up her heart to him and soon they go into business together." He sat down on my couch and grabbed my remote control. "Then it turns that he had actually raped her several months earlier," he said. "That's the bad experience I was talking about."

"That's supposed to be inspiring?"

"Yes, because in the end, she kicks his ass back to jail."

"Look, I've decided to embrace being single, so I don't need cheering up," I said, settling down next to him.

He rolled his eyes. "You? Right."

"I'm serious."

He shook his head. "I don't know why you're so terrified to be alone. You're actually calmer when you're alone. Whenever you're around people, you're like a cat inside a carrier on his way to the vet. 'Frantic' is the word that comes to mind."

I sighed. "I just get depressed at the notion of spending each day surrounded by nine million New Yorkers and then going home to my little box. I just feel like I'm in a kennel."

He jumped up to retrieve the box of candy and cracked the cellophane. "Maybe you should try to actually enjoy the solitude," he said, inspecting the diagram to locate a vanilla cream. "Maybe you should entertain the possibility that you might be alone for a very long time. Maybe forever. I'm not saying that's going to happen, but it's better than cancer." He shrugged. "You could be with Trevor. Why are horrible relationships so much better than being alone? I've been in bad relationships and now I'm single. Single's better. Sometimes I'm a little depressed because I know what it was like to be with someone I cared about, but the older I get, the less depressed I get, because I really don't see that many relationships I'm envious of, anyway." He held out the box to me. "You've got to move forward," he said. "You've got to find some other reason to be happy, just because you need that, anyway."

On weekends, I started to leave my apartment early in the morning with no agenda. All day I would roam Manhattan, stopping whenever something caught my eye. I spent hours in museums and bookstores and found that I liked my own company.

As I headed out one Saturday morning for a long jaunt (breakfast, an early movie, an exhibit of Lewis Carroll's photography, maybe a swing by the farmers' market at Union Square), my father called.

"Hi, kid," he said. I was becoming the world's oldest kid. I felt like the forties radio act Baby Snooks. "Haven't heard from you in a while."

"I know, Dad. I'm sorry."

"Listen, I've been thinking. I'd like to invite Trevor for dinner at the house."

"Too late," I said. "I broke up with him."

He exhaled in a whoosh. "Thank God," he muttered.

"Let's not get into it," I said. "I know he wasn't right for me, but I guess I just had to do it. It's the only way I'm going to learn what I don't want."

"Well, maybe we shouldn't have said anything. Over the years, you've basically proved that your instincts are right and maybe we should trust them more." He sighed. "We've always been careful not to oversteer you girls. We kind of had to let you bruise yourselves and make mistakes. That's what my parents did."

He paused. "I don't know half the bruises you've gotten, but on the face of things, it seemed that you were doing fine and handling it, and I guess I was hoping that you were. And we think about things in our generation's terms, unfortunately. If I had my way when it came to your career, you'd still be at that ad agency. To me, working at *Rolling Stone* did not sound like a good move, but we didn't want to control your life. And you were right. So I'm sorry if I stepped on your toes with that guy. I just want somebody for you that's challenging. I always felt you were artistic, and if you had a real stable partner to ground you, then you could explore your artistic side, which would be terrific." He laughed. "I know you don't want to date anybody like your old dad. I'm sure you want a snazzy New Yorker, not some quiet guy from the Midwest who likes vanilla ice cream with his pie."

"You know what, Dad? I'm staying single for a while."

"Sounds good to me," he said. "Say, what are you doing today?" My father really did use the word *say.*

I thought for a minute. "Well, I kind of have plans."

"Well, break them."

The two of them must have jumped right into the car, because a few hours later, there they were in my doorway, my father holding his tool kit to make any minor repairs in my apartment, my mother holding a small stuffed rabbit wearing a hat and glasses.

"I can't get away from you people," I said.

"Nope," said my father.

"Your father says you broke up with Trevor," said my mother, grabbing my arm. "What happened?"

"Jesus, Ma," I said. "Shouldn't you soften me up first with a little small talk?"

The Difficult Question: When to Bring Up the Church of Scientology

There comes a point in every interview when the awkward question must be posed—the one about your star's bitter divorce, or alarming weight gain, or extended hospital stay due to that old euphemism "exhaustion." Alternately, there is the question that will never go away, the one that will be asked in perpetuity: Kate Winslet will always be made to discuss her nonexistent weight problem, Jennifer Lopez will forever be asked about her *tuchis,* George Clooney cannot elude questions about why he won't "settle down." These tired perennials never fail to irritate your subject, and rightly so, but there is, for reasons that mystify, an insatiable demand for endless variations of the same answer.

There is no method to erase the dread of asking the awkward question, but there is one way to at least minimize the damage, should your star become incensed. Remember above all else that if you are in a restaurant with your subject (and as I have pointed out, 95 percent of the time, that is where you will be), *get the check and pay for your meal before the question is asked.* Otherwise, if you flame out, that wait for the check will be long. And tense.

Every interviewer goes about this process in the same way. After you have safely signed the bill (do not attempt to ask the question if the waiter has just taken your credit card, because God only knows when he might return and it's just too risky), affect a sheepish, self-effacing demeanor. Then say, "I certainly don't want to pry into your personal affairs, which are absolutely none of my business (light chuckle), but my editors wanted me to ask about your eating disorder/felonious brother/lip augmentation, even though (bemused headshake, derisive, empathetic snort) I don't really understand when it became the public's right to know this stuff."

It is useless to try to wriggle out of the awkward question, because prior to your interview most magazine editors will send you an e-mail in all capital letters saying ESSENTIAL TO THE PIECE, even though in many cases, they are questions that you wouldn't ask some of your closest friends. How'd the abortion go? Why, exactly, are you and your husband getting divorced? Was it the whole hooker thing?

In a different setting, many people would throw down their napkin in disgust and stalk off at this sort of prying, but not in this particular zone. If you have softened the person enough, or if they're still new to the game, they may offer up a personal anecdote. Or they may provide a general comment, which will at least result in your getting paid. You simply preface the parsimonious quote with a wordy, inflamed lead-in: "Of the shocking charge that she had her housekeeper act as her drug runner and subsequent grim stretch in rehab, she says simply, 'I just want to move on with my life.' "

This sends a message to your editor: *Hey, I tried. Me, I'm not scared to ask the tough questions. It's not like I can physically force her to go into detail.*

My check-grabbing lesson came courtesy of *Flashdance* and *The L Word* star Jennifer Beals. At that point, I was pursuing in earnest more work at women's magazines. Doing profiles for them was wonderfully pleasant. After years of dealing with hungover rock stars, I just wanted people to behave themselves, and to my profound relief, most of my subjects were affable female country music or sitcom stars, all of them roughly my age. I'd fly into Nashville or Los Angeles, have a just-girls chat on how they balance work and family and what bad habits they wish they could break. Then I'd fly home and write the piece. Easy. Painless.

My favorite was *Lifetime* magazine, a print extension of the network that lasted two years, during which time I had coffee or brunch with reliable sellers like Faith Hill or *CSI*'s Marg Helgenberger. "Couldn't be nicer," I would inevitably report back.

Then came Beals. She had a reputation for being a prickly interview, but surely, I thought, not for soft, friendly *Lifetime*. Beals, a well-traveled Yale graduate, just didn't play the anecdote game—which I would have completely respected if I hadn't had a job to do—and was known to be incredibly guarded about her personal life.

I met her at a restaurant in Santa Monica. Tall and serene, she glowed with yogic good health. She was wary but cordial—until, that is, I asked her about her husband.

"I just have to throw in one question about him," I said apologetically. "Maybe something about how you met him."

Her tight smile vanished, her eyes narrowed, and she snapped in a loud voice that she didn't *have* to do anything, and no, she was not going to answer. As she went on, her voice rising, a concerned waiter hovered nearby.

There was something surreally appalling about being dressed down by Alex Owens: Pittsburgh welder by day, exotic dancer by night. I tried to maintain my composure after she calmed down, but the rest of our lunch was terse and uncomfortable, made more so by the interminable wait for the goddamn check.

I was still rattled when I arrived back at my room at the ritzy hotel that I had wrangled. As I closed the door and slipped into a robe, I was horrified to burst into tears.

"I can't believe Jennifer Beals is making me cry," I sobbed to Dinah on the phone.

"Who the hell is she?" Dinah said. "She was in a dance movie and she can't even dance. Is something else bothering you?"

Celebrities were bothering me, and while I loved my fancy hotels, a creeping lonesomeness always set in by the second day. Some of my friends with kids couldn't wait to travel for business to get away from the tumult at home. I didn't have any tumult at home.

After I hung up with Dinah, the phone rang again. It was the folks, whom Dinah had obviously alerted.

"Hi, kid," said my father.

"We heard what happened," chimed in my mother on the bedroom extension. "Stupid bitch."

"Too bad," said my father. "She was really good in that movie where she had cancer."

I stopped snuffling. "What?"

"The one where Shirley MacLaine was her mother. Remember?"

"Jay," said my mother, "Jay. That was *Terms of Endearment*. This was the one who was in *Flashdance*."

"Oh," said my father. "Well, she was good in that, too."

Someone was knocking on my door. "Folks, I've got to go," I said.

I opened the door to a maid. "Do you want turn-down service, miss?" she asked.

"Sure," I said, ushering her in. It was embarrassing to have her turn down the bed while I was perfectly capable of doing it myself, but I wanted the chocolate that she put on the pillow.

She looked critically at me as I blew my nose. "Are you okay?" she asked.

"Me? Oh, sure," I said, throwing the tissue away. "I interview people for a living and I just had a bad experience. I don't know why I'm upset. Some people get snapped at every day."

She plumped the pillows with a neutral expression, but I saw her eyes flick toward me.

"It was Jennifer Beals," I said.

"*Flashdance,*" said the maid, whose name was Ana.

"Right."

She turned and looked at me. "Famous people," she said. "Some are nice, some are so crazy."

I stopped sniffling and suddenly felt ashamed. "I imagine you see a lot of strange behavior," I said.

Ana pulled back the comforter on the bed and expertly folded the sheet.

"Oh, honey, you don't even know. And we can't say a thing. We just do our jobs. Mariah Carey was here a few weeks ago with bodyguards by the door. I was sent up there to clean the room, but they wouldn't let me in," she said. "I didn't know what to do." I pictured her hesitating in front of the door with her cart.

Ana reached over and patted my hand. "Don't you worry about it," she said, turning to go. "She doesn't know you. That's what I tell myself: They don't know me."

A half hour later there was another knock at the door. It was Ana again, accompanied by one of the bellhops, who carried a towering basket of fruit and chocolate.

"I told him what happened," she said, "and we thought you might like this." Somehow they had finagled it from Guest Relations.

Of course, I burst into tears again. I invited them in, and we spent fifteen minutes trading celebrity war stories—theirs far more gruesome because, as the bellhop pointed out, "they don't see us, so for some reason they think we don't see them."

After we were through commiserating, Ana glanced at the clock by my bed. "We should go," she said, making for the door. "Hope you feel better."

When I got back to New York, I vowed to make a quick getaway in future interviews, after potentially explosive questions. I also sent Ana the biggest bouquet of flowers that I could find.

16.

I had just walked into my apartment with my nightly serving of dinner from the gourmet place when I got a call from Julie to tell me about her first day at VH1. Relations between us were almost back to normal. Neither of us mentioned the rift again, and we had faked that it never happened so effectively that most of the awkwardness had dropped away.

"How did it go?" I said, throwing on my favorite pair of saggy-drawered sweatpants.

"Do you know how I have been e-mailing back and forth with Paul, my boss?"

"Right, right," I said, feeling under my bed for my slippers.

"He's the head writer and producer, and I knew that there was some sort of attraction because our exchanges were very funny and flirty. Plus, I was so nervous because I had never had a job except at the insurance company and he was really comforting and promised he would help me."

"Good," I prompted.

"Well, the minute I met him today, I knew there was something. I can't explain it. He just looked very familiar to me."

"How so?"

I heard her pouring kibble into a bowl for her dog. "I had the pang that Cher said she had when she met Sonny and Rob Camiletti. She says that the time that she met those two guys, the rest of the room went dark. That's what happened when I met Paul." I had never heard her sound so excited. Julie was a chronic dater, while I was a chronic inappropriate-boyfriend collector.

"So listen, I need your advice," she continued. "There is this party after work tonight to celebrate some special that Paul produced, but I went home first because I was exhausted. When I worked at that insurance company I was used to quitting time being at, like, three, and today we worked until six thirty. So I told him that a window had broken in my apartment and I had to take care of it but I'd try to go later. Okay. So. He just left me this message."

"Play it."

She held the phone up to the machine. I heard a slightly faltering male voice say, "Hi, this is Paul. I hope everything went okay with the broken window, and I hope you can make it to the party later, because I . . . well, because I was really looking forward to getting to know you."

"That's not a 'howdy, coworker' sort of message," I diagnosed. "He's clearly into you. Be careful. You just started this job."

"I know. But I should go to the party, right?"

"Absolutely."

She snuck a call to me the next day at work. "I did go to the thing," she said in a low voice. "I sat next to him, and the other producers were telling funny stories and Paul was telling me about the time in his life that he was the most depressed."

"So then it didn't go well?"

"Actually, it did. When he wasn't telling me that story he made me laugh like I haven't done in years. Like almost wetting my pants."

A few nights later, they went on their first date at a downtown tavern, and I waited anxiously for her morning report.

"Are you ready for this?" she said, calling me before I had even poured my cereal. "This morning I told my aunt Mattie and my aunt Phyllis that this was the guy I'm going to marry. He's so kind. He makes me laugh. He has the same name as my father and they have the exact same birthday, thirty years apart."

I was floored. This sort of hyperbole was completely unlike Julie. "Do you really think you're going to marry him?" I asked.

"How the hell do I know?" she said. Then she laughed. "No, I really do think I will."

"Oh, Jul," I said. "This is the happiest news."

"How about you?" she said. "Maybe it's time for you to get out there again."

"I don't think so," I said. "First of all, it's not like I've been besieged with offers. The only prospect is that setup that Casey wants me to do and I'm not going on a blind date." My friend Casey, a publicist, kept trying to push a writer on me named Tom.

"He's handsome, and he writes for the *New York Times,*" Casey had wheedled. "Your favorite, the City section. He plays soccer. He's incredibly smart."

I asked my time-honored question. "Would you sleep with him?" I demanded.

"She says that she would sleep with him," I told Julie. "I mean, not that she would, but theoretically. For weeks, I've been putting her off, but she's wearing me down."

"Oh, just go," Julie said. "I always told myself that it's just one night. Who cares? If it's bad, you'll have a funny story to tell me afterward."

I crawled into my bed. Maybe I would bring dinner in here on a tray, like some sort of old film star, and eat while I flipped through magazines. "True," I said.

I called Casey a few days later. "Okay, I'll do it," I said. "Only if you invite someone else so that it doesn't look like a complete setup."

She told me to meet her at a soul food place in the twenties called Lola's. When I walked in, I realized that I had been there before, during a gospel brunch. I had a fleeting recollection of resolutely eating my peach cobbler as a group of singers pulled my mother onstage to sing background on the hymn "How I Got Over."

Casey had brought along her boyfriend and another one of our friends to act as a fifth-wheel decoy. I greeted everyone, ordered a sparkling water and cranberry mocktail, and took a seat, fiddling with my silverware and trying not to stake out the doorway.

Ah. There he was. I could tell by the uncertain way he scanned the room. As he approached the table, I turned discreetly to Casey. "No," I said.

She shook her head. "You are ridiculous."

I leaned closer. "I can tell you right now that he's not my type," I said quietly. For starters, he carried a battered leather satchel straight out of *Goodbye, Mr. Chips* that screamed "Hey, I'm a writer!" And he towered over me at six foot three while I had always seemed to end up with men who were more eye-level. He had the neat, crisp look of a man in a fifties-era shirt ad: cleft chin, blue eyes, slim, with short, sandy brown hair. He seemed so familiar. Why was that? Then it hit me: He looked exactly like the jazz trumpeter Chet Baker, before the drugs.

We all started off with a little general conversation, and then I figured that I should humor Casey and talk to Tom a bit. The problem was that he sat across the table from me, so every time I addressed him, I had to raise my voice.

"So Casey tells me you're working on a book," I said loudly. Immediately our three tablemates ceased talking and leaned forward. They obviously wanted to see if we were getting along. I felt like we were pandas at the zoo, and everyone was breathlessly monitoring us to see if we were going to mate or not.

Tom cleared his throat. "Well, ah, yes, it's sort of a . . . I don't know how to describe it." He tried again. "It's a survey of Cold War landscapes."

I rummaged my mental archives for some sort of question, or even a relevant comment. Hm. Nothing.

"That sounds dry." He corrected himself. "I mean it's a travelogue of—well, you know, missile silos and relocation bunkers across the country. Atomic proving grounds. That kind of thing."

"Ah," I said.

Nobody uttered a word. Where was the waitress? I prayed that no one would order appetizers. If we all went straight for the entrées, a good twenty minutes could be shaved off the evening.

Tom leaned forward in a vain attempt not to shout. "And I understand you write for *Rolling Stone,*" he said. "Who have you interviewed lately?"

This, I could do. "Let's see," I said. "Yesterday I talked to a guy from Saliva."

"Saliva. Saliva," he said. "I only like their earlier work. No, I don't know them. I was a deejay in college but I'm afraid my knowledge pretty much stopped at that period. I could talk about the Descendents, if you want."

I smiled. "Believe me, I'm happy not to talk about music. I'm in the midst of a career change. I was old enough to be the Saliva guy's mother, let's put it that way. So you write for magazines as well?"

He nodded. "Mostly design magazines. I write about architecture, history, things of that nature. I just wrote a piece about the cultural history of the revolving restaurant."

"Aren't they all history? They seem sort of stuck in the seventies to me."

Tom sat up, invigorated. "Actually, there has only been one built in the U.S. in the last ten years, but there has been a spate of new ones built in the Gulf-Arab states . . ."

As he went on and on, I began to feel like I was in a revolving restaurant myself. Too bad. He was nice, which was, admittedly, a notch above Perfectly Nice and a few notches above Fine, but we weren't clicking. By the time our food arrived, I noticed that our tablemates had stopped their eavesdropping and had resumed their conversation.

I snuck glances around the room as he moved onto another article he was working on about the unusual origins of everyday objects. "The Dixie cup was created as part of a public health campaign against mass viral outbreaks," he was saying. "And oh! Here's something: Bubble wrap was the accidental invention of two scientists trying to create plastic wallpaper . . ."

Tucked away in a corner, I noticed, was a tall woman seated at a little table next to a sign that said READINGS BY NEFERLYN. On the table sat a burning candle and—I had to subtly crane my neck—a deck of cards.

Tom and I moved on to the "biography" section of our setup. "So why do you live in Brooklyn?" I asked.

He cycled through the reasons—it's cheaper, more space, no tourists, more trees—and then I offered a defense of my own neighborhood with my "Murray Hill isn't so bad" rap. Dessert arrived.

I stood up. "I'm going to visit the fortune-teller," I announced to the table. "I feel badly that she's not getting any business."

I drifted over to her table. Neferlyn was a tall, queenly woman with blond hair who wore an appropriately dramatic, sparkly purple dress. She had the perfectly penciled eyebrows of a thirties-era actress. She wasn't immediately identifiable as black, or white, or male, or female. She seemed to be a little of everything.

She gazed unblinkingly back at me. "I accept thirty dollars," she said.

"Oh," I said, fumbling in my purse. "Of course." I put the bills on the table and in a smooth, discreet motion, she tucked them under the table somewhere.

"Let's do your cards," she said, doing a few shuffling moves before methodically laying them down on the table. "Mmm," she murmured.

"Mm?" I said. "Is it something bad?"

"No, no. It's interesting. There are children in your life . . ."

Nope. Oh, well. Good-bye, thirty dollars.

She frowned and studied the cards. "But they are not yours."

"One of my sisters has a baby, and the other is due shortly," I said. She pored over the cards, her lips moving slightly. "Listen," I said hesitantly, "what if you did see something bad? Would you tell a person?"

She allowed a small smile and said that yes, she would find a way to warn them if it was serious, but no, she wouldn't scare a person to death.

As she considered the cards, I couldn't resist quizzing her. How long had she been at the restaurant? Did people ever come over when they were drunk? Do you have security in case they get out of hand? Oh, most people behave themselves? Huh. Interesting.

Neferlyn drew herself up to her full height. "We contain within us all of the answers to the situations that challenge us," she announced. "I see that you are in a transitional period," she continued, her eyes moving over the cards. "This year has been difficult for you," she said. "Someone has left your life who was very dark. Very bad for you."

I leaned forward. "True."

"I see here that you are going to travel," she said. This did not impress me. With airline prices at record lows, who doesn't travel?

"I see also that . . ." She turned over another card. "I see that you are going to meet a man from the Midwest who will give you his heart."

"Well, everyone I meet is from New Jersey or Long Island," I said. "Maybe you mean my dad. He's from Michigan."

I finally got her to laugh.

"He will give you his heart. I can only tell you what I see."

I stood. Time to return to my table and wrap the evening up quickly. "Thanks, Neferlyn," I said.

"I'm available for private parties," she called as I made my way back to the table.

The Gold Standard: Successfully Worming Your Way into the House

If you are actually invited to conduct an interview inside a celebrity's house, jump on it. This is the bonanza of profiles, because your work is essentially done for you, especially if you're sent to the home of a famous person who has the good sense to realize that they have a civic duty for their decor to be over the top. We regular folk do not want to see a Crate and Barrel sofa, a hair-encrusted dog bed, a non-plasma TV, and a squalid pile of newspapers. We want glitz. We want to see something that we do not have in our own homes.

Star Jones knew this. Her plush Upper East Side apartment in New York was exactly what you would expect of a glamorous TV personality. An assistant opened the door, swiftly glided away with my coat, and returned with a cold beverage. The first sight that greeted me was a capacious oil painting of—who else?—Ms. Jones herself, which hung in the stairway. Every celebrity should have a giant oil painting of themselves in their house. I frankly feel a little insulted if there isn't one. What's the point of being famous if you don't have a likeness of yourself that rivals the one in *Laura*?

Everything was plush, abounding in gold and cream tones, accented with animal prints, and there was a large spread arrayed on the gleaming dining room table: fruit, crudités, meats. It was all assembled by Star's personal chef, who even had the appropriate name: Bianca. Who wants something prepared by Joanne? Ah, yes, thanks so much, Bianca. Champagne was offered. The lights were dimmed. In swept the appropriately named cohost of *The View*. She was wearing her off-duty clothes: a pink rhinestoned tank top and velour sweatpants combo with a fluffy tan robe. Who doesn't love a girl in sweatpants and full makeup? Diamonds glinted from her ears, fingers, and toe ring as she happily nestled into a couch and got the ambience going by grabbing a remote control to click on video footage of fish, which flitted tranquilly around her enormous TV screen.

Together, we paged through an album of photos of her recent trip to the Bahamas with her fiancé, Al Reynolds, and some friends ("What a good-looking group of black people!" she had scribbled on one). Then, after a chat, it was time for one of the most beautifully thorough house tours I've ever had. "I've got a bidet!" she cried, opening the bathroom door. "And look at this shower! Ten people could fit in it!" Another glamorous photo of her rested on the bathroom counter. As, I might add, it should. On to the office, which had spangly gold stars on the walls and a chair with a large gold S stamped on it, and then the bedroom, where she opened up a couple of drawers to display her towering pile of Louis Vuitton purses. If only every famous person opened up their drawers for you, what a wonderful world it would be.

"You want to see the closet?" she asked unnecessarily, flinging it open. Rows and rows and rows of shoes lay in the light-up, floor-to-ceiling closet, and yes, there were plenty of Payless among the Manolos and the Jimmy Choos.

We trooped downstairs, where her assistant awaited. "The gift bags are here from last night," he said. "Shall I load them in?" Load? What was in those things?

She grinned. "You know it!" she said with relish.

Being in her house was perfect for the story, because then you could see a person who worked hard for her success—in her former life, she was a Brooklyn D.A.—who was enjoying the fruits of her labors, every cut-up piece.

A home studio can work well as a setting if it is sufficiently large enough to express an outsize personality. Kid Rock's studio, part of his home compound situated about an hour's drive outside of Detroit, was custom decorated to be a NASCAR fan's idea of paradise. Rock led the house tour through the gleaming garage, which boasted some real Vegas slot machines, a fridge covered by the flocked industrial steel that truckers use, and a wall-mounted TV with built-in speakers facing the requisite bachelor black leather couches (with cupholders, natch). Parked right in the middle of the garage was a huge, lustrous motorcycle as well as the orange General Lee car, of which only a handful exist, driven by—cue choir music!—the Dukes of Hazzard.

In the kitchen, sustenance was provided by one Costco-sized vat of pickles. All over the building, vintage record covers were tacked to the wall: ZZ Top, Merle Haggard, and, adding a note of pathos to the proceedings, many tributes to Joe C., Rock's deceased dwarf pal: photos, plaques, posters. On to the basement, which housed a bar with neon signs (just what a twelve-year-old fantasizes that his den will look like), a guitar shaped like an Airstream trailer propped against the wall, and, the pièce de résistance, a small stage with lights featuring a stripper pole. It wasn't shiny, either. Its surface was a tad opaque. From use. My job was done.

Certain stars have such a unique persona that you just know their house will follow suit. When Stevie Nicks invited me to her Los Angeles home, it was suitably magnificent. Garlands of red silk roses snaked up the walls and lined the ceilings. Fringed shawls were strewn over every available surface—on the piano, as a tablecloth. A massive painting of a big-eyed gypsy girl hung in the living room near the piano. Stevie's taken a lot of grief for her Tolkein-ish nightbird image—the spinning skirts, the white-witch imagery—so she was guarded at first, but when she saw my irony-free rejoicing in her décor, she began to warm up.

It was no act. I loved her unabashedly. One of the sturdily reliable questions that I ask musicians is "What song will put a lump in your throat, no matter how many times you have heard it?" Well, for me, it was "Silver

Spring," a sorrowful rebuke to a man whose interest is waning. I could barely listen to it without fumbling for a tissue and thinking *God, why couldn't she and Lindsey Buckingham make it work?*

Barefoot and in a demure floral sundress, she padded up the stairs to show me her clothing-stuffed dressing room, which was lined with suede platform boots in every color imaginable, like a groovy box of crayons. And, joy of joys, she let me try on one of her shawls! "And these beads belonged to Janis Joplin," she said, putting them on me. Heaven!

Stevie had a squad of young blond assistants who were stationed in various parts of the house. "Want to see my new video?" Nicks asked at one point, but she couldn't work the VCR. "I can't—how do you . . ." she flapped helplessly, trailing off, before an assistant rushed in. Another set out lunch, a variety of healthy salads, on the kitchen island. Then Stevie and I sat close together on her couch as she told me—in gloriously uncensored detail—about her former coke addiction and tortured romance with Lindsey. Interviewing people who came of age in the sixties or seventies is so much more rewarding than talking to today's bland, p.r.-schooled youngsters. During one week, I chatted with Justin Timberlake and Grace Slick. Timberlake, so cautious, so eager not to offend, weighed and measured everything he said. As a former Mouseketeer, he was trained from a young age in how to handle the media. As a result, he was pleasant, but mostly stuck to safe fare such as how it's not about the fame, it's about the work, and his appreciation of his fans, and that being on the cover of *Rolling Stone* was really cool. Slick, meanwhile, cheerfully talked about how she couldn't fully participate in an orgy that sprang up in Jefferson Airplane's San Francisco office because she wasn't good at multitasking, added unapologetically that her lungs were "two black bags" from smoking, and mused that her only regret in life was that she never nailed Jimi Hendrix or Peter O'Toole.

But back to Stevie. After we covered the coke and the romances, she brought out her velvet-covered, poetry-filled diaries from 1979's *Tusk* tour, when she had started an affair with drummer Mick Fleetwood, and we read them together, sometimes aloud.

After a few happy hours, Nicks and her rambunctious assistants broke out the binoculars to spy on the neighbors. "You can stay over, you know," she said to me. "There's a fabulous guest bedroom." She showed me the impossibly high guest bed, which had a dramatic red drapery hanging from the ceiling. They all planned to make a little dinner, play some music, watch *Golden Girls* reruns—really, up my street in every way. I was tempted, but I thought I should maintain a professional boundary and declined. How stupid was I?

I learned my lesson when I was invited to the Tennessee ranch of Loretta Lynn. If she asked me to stay, by God, my bags were packed. Sadly, she didn't, but I did get to spend a long afternoon making peanut-butter fudge with her in her kitchen.

When you roll up Lynn's driveway—overlooked by a house on a hill inhabited by her oldest son, who likes to watch over his mama—a sign was posted that said NO TRESPASS'N. The first thing you noticed was a vegetable patch, because Lynn still grew and canned her own vegetables. Parked in the driveway was a two-toned purple bus with LORETTA LYNN written in cursive on the side.

Lynn greeted me with a big kiss on the mouth, which didn't faze me in the least. Having read her two autobiographies, I pretty much knew where she had been. (Her first kiss came courtesy of Doolittle Lynn, the man who would end up being her husband of forty-eight years.) In fact, I was touched that she gave me a kiss, not knowing where *I* had been. She was wearing black pants and a purple shirt with rhinestones. No shoes, just those black knee-high nylons that some ladies wear instead of socks.

She was so welcoming that I instantly felt at home, and because my mother's family is from Alabama, the whole scene was all very familiar, starting with the preponderance of sweets. Arrayed on the kitchen island were a cake under a glass container, a tray of brownies, and another of no-bake cookies (a tasty mixture of oats, peanut butter, brown sugar, and butter).

Usually, celebrities do not think to feed you during an interview (I always toted some variation of food that you find in a birdfeeder: nuts, seeds, and dried fruit, just in case), but Lynn offered up chips and salsa, some coffee—dark and sludgy, the way she likes it—and the best part: bologna sandwiches. She pushed some homemade bread over (she bakes a loaf every week) and pointed toward the fridge, urging me to help myself to the bologna. The door had a magnet on it that said IF MOMMA AIN'T HAPPY, AIN'T NOBODY HAPPY. Tentatively, and then with more assurance, I opened the fridge and rooted around.

Then came the tour of the house, which was built by her husband, Doo, who died in 1996. There were dozens and dozens of dolls all over the place—antebellum dolls, some of them waist-high, with frothy, candy-colored gowns, and Indian dolls (Lynn is part Cherokee). Often, people who grew up poor are comforted by the presence of lots of stuff, and Lynn was no exception. Thus, there weren't just a couple of baskets hanging from the kitchen's ceiling; there was a flotilla of them. On every available surface were Native American dream catchers, ceramic flowers, and plaques with inspirational sayings on them.

Gifts from fans were everywhere, because the tenderhearted Lynn couldn't bear not to display them. In one room, there was a crocheted afghan draped over the couch that said COAL MINER'S DAUGHTER, made by an older lady in a wheelchair. Lynn's assistant, Tim, showed me another present that was housed in a glass case. It was hard to tell what it was. It looked vaguely like lollipops in a jar. Further scrutiny revealed that it was a bouquet of "flowers" whose petals were made of plastic spoons in different colors. In a more formal room, there was a large oil painting of Loretta done by a guy in prison.

Almost as fun was a tour of the Loretta Lynn Coal Miner's Daughter Museum, which was also on the grounds, along with a replica of the cabin she grew up in, a campground, a racetrack, and a country and western wear store. The woman saved everything (she even had Tim save and dry the flowers that fans gave her, for potpourri), so the museum had a display of her fabulous gowns, Doo's old Jeep, her daughter's report cards, and even some presidential memorabilia, including a pair of immense yellow pumps donated by Barbara Bush that looked like something a trannie would wear.

When it was time for our sit-down, we ensconced ourselves on the couch, shoveling down chips and salsa while she talked about her old friends Patsy Cline and Tammy Wynette and life in her birthplace of Butcher Holler, an isolated area in eastern Kentucky. As a kid, she went without shoes, didn't see a toilet flush until she was thirteen, and often ate bread dipped in a makeshift gravy of browned flour and water. She was a link to a time and place that seemed incredibly remote, and the forces that shaped her as a child do not exist anymore.

She was a born storyteller—which is why the songs that she wrote were so masterful—so it was easy to wind her up with a few key words while we repaired to the kitchen to make peanut-butter fudge. She bustled around, directing me to get her the butter and to dump some nice big globs of Jif into the bubbling pot. I wanted to stay the weekend. You could ask her anything and she would answer. Her publicist had said that Lynn had no filter, and it was true. After a while, I asked her about a chapter in her first autobiography, *Coal Miner's Daughter*, in which she talked about getting pregnant at thirteen. She had written that when the doctor told her she was pregnant, she didn't know what the word meant. She was similarly mystified when she first got her period, and ran down to the river and jumped in to get rid of the blood. She thought that she was going to die.

She talked so much that we neglected our candy, and it ended up hard and crumbly. "My pot done gone to pot," she said. Unfazed, she stuck some spoons into the pot and brought it, still warm, over to the table, and we gleefully dug in.

A few days later, when I had returned to my New York apartment, a package arrived in the mail. It was from Loretta. It contained a gingham apron and a big block of peanut-butter fudge. Also a note: "I wanted to send you some to show you how it's supposed to be!" she wrote on LL-emblazoned stationery. "Love you, Loretta Lynn."

17.

After my chat with Loretta I decided it was time to change course in my career. Freed from *Good Morning America* and MTV2 and reaching Old Veteran status at *Rolling Stone*, I vowed to be more judicious in my choice of interview subjects. It was time to stop grabbing at every assignment. Instead, I would just take on people with whom I could really communicate, or ones I truly admired. I was getting weary of being that mechanical monkey, clapping my cymbals together and trying to ingratiate myself.

"That's why I liked Loretta so much," I said to my mother one night as we dried dishes in her kitchen. I had traveled to their house from New York for "Steak 'n' *60 Minutes* Sunday." "Loretta's house reminded me of Aunt Eunice's house, a little bit. I identified with her, in a small way."

"Well, I think that's great," said my mother. "Good for you. You don't need to talk to every sitcom star that comes down the road. Enough, already."

"Get this," I said. "Loretta used to eat fried squirrel."

My mother put a platter on the counter. "Well, what the hell do you think I used to eat? Dad would take a shotgun out into the woods."

I stared at her. I spent days feverishly preparing for some of these interviews, poring over celebrity research, and the irony was that I had a rich trove of stories right in front of me. But who thinks to interview their own mother? As a self-fixated teen, I never imagined that she had an actual personal history. To my young eyes, she was Source of Cash Obsessed with De-Cluttering. After all of my preparation, I could recite Loretta Lynn's background from birth onward, but there were glaring gaps in my own family lore. Could I write a profile of my mother? I wasn't sure. Shamefully, I realized that I hadn't even been curious enough to ask her the celebrity questions that I used in case of emergency.

I decided to pull out one of my standbys for women's magazines. "Hey, Ma," I said. "What makes you sentimental?" I was honestly curious. Overly emotional she was not. She was many things—funny, bright (she drove us kids to the library every week and made us pick out three books each, fostering a lifelong love of reading)—but my father was the hugger. She was the no-nonsense parent who briskly marshaled things along. Although I noticed that sometimes a glass of Pinot Noir could produce a little fissure in that coolly capable veneer.

I poured her some wine and sat down at the kitchen table, pushing the glass toward her. "I can't remember the last time I saw you cry," I said. This was bold territory. Normally I didn't venture into the self-reflective with her, but now that I lived a whole hour and a half away in New York, I felt that we could chat as two adults.

"Well, that's how much you know," she said. "I cried last week. Thinking about Dad." Her favorite picture of our grandfather hung in our living room: Hershal Ray Corners, his skinny butt perched on an overturned bucket as he fished in a muddy stream near their hometown of Citronelle, Alabama. Small and wiry and universally respected, Hershel was the office manager of the Gulf gas station in town. He taught my mother to drive when she was thirteen and determined not to show she was terrified as the car barreled down the town's dirt back roads.

She laughed and took a sip of wine. "Dad's word was the law," she said. "He was small but he had an incredible presence, so nobody went against him, including Mama. But he had a soft spot that he only showed occasionally." She smiled, preoccupied. I thought maybe I should prompt her, like I did with famous people.

"Maybe you two are alike, in that way," I offered.

"Maybe we are," she said. "I was never afraid of him, but what he said went. I remember when I was a high school freshman, I went to the skating rink. When Dad went to pick me up, I had already left to go to the drive-in with an older boy. He was sixteen. You can imagine the horror I felt at the drive-in when I saw Dad's car cruising slowly between the aisles, with just his parking lights on." She laughed. "It was like a shark. We immediately left. When I got home, Dad was furious and I was grounded for a month. But within two days he relented. You know, they looked the other way because I was the baby."

She hesitated again. My years of indifference had probably made her mindful of talking too much. Oh, the shame. "Hershal loved to fish, didn't he?" I asked.

"Oh my God," she said. "We used to get up at three in the morning so we could get there when the fish were biting early. My brother Bill told me once that they were fishing on the Tom Bigbee River, and it was getting dark, when they got into a huge mess of fish, and they were pulling them in one after another. Finally Bill said, 'Come on, Dad, we have to go.' And Dad was saying, 'Just one more. Just one more.' And he was actually holding on to an overhanging branch so that Bill couldn't pull away in the boat." We both laughed.

"I remember that you used to go deep-sea fishing on the Gulf," I said.

"That was kind of an unspoken test in our family, that when you went fishing on the Gulf, you did not get seasick," she said. "It was a matter of pride, I guess. The only time I ever got seasick was when your father and I were first married, we went with Dad way out on the Gulf, and a squall came up, and that boat was rocking and rolling." Unconsciously, her southern accent was thickening. "Your father and I went below with everyone else. We were literally green. I looked through the porthole, just to see who might still be fishing. And there was Dad, with his feet braced against the railing, his line still in the water, and in the other hand he was clutching onto this greasy fried egg sandwich." She laughed. "He was tough."

"So what were you crying about last week?" I asked. "Were you just missing him in general?"

She sighed. "Oh. Well, I was thinking about when he died. It just comes into my head sometimes. Mother had called me home, because Dad was in the hospital and was dying of lung cancer that had metastasized to his brain. I left you two older girls at home and I took Heather, who was, I guess, five, to Citronelle with me. I think we got there on a Thursday. The hospital was very small, it was in Chatom, Alabama, where I was born. Everybody knew everybody, all the doctors and nurses."

She sat quietly for a moment. "Dad was unconscious, we thought," she said. "We didn't know if he could understand us, we didn't know if he could hear us or not. But I decided to talk to him, telling him what was going on, just in case he could understand." She smiled. "Since he was so interested in you grandchildren, I told him about Heather's latest adventure at the pediatrician at home. I had taken her there and we were a little early, and the waiting room was deserted except for a very old horse-faced woman with a cane, who kept it squarely between her legs. Heather sat in the chair next to me and I remember I picked up a magazine and got so involved in an article—I think it was about gardening, or something—that I really didn't notice what was going on. When I looked up, Heather was not sitting in the chair any longer but had somehow finagled the cane away from the lady and was doing a soft-shoe dance. Jumping around, like some sort of vaudeville act."

My mother threw back her head and laughed. "She was so happy. And the old horse-faced lady was silently laughing her head off. Never made a sound, with her hand up to her mouth covering her teeth! I thought that was one of the funniest things I had ever seen in my life."

She distractedly wiped at a tear that slipped out when she laughed. "And when I told this story to Dad, he actually smiled. His eyes were shut, but I knew he could understand. That he was there. So that redoubled my efforts to talk to him and tell him everything that was happening."

I waited, motionless. "And then the next day," she went on, "for the first time since he was in the hospital, Mama didn't go to visit him, and she kept Heather at home with her. It was raining, and I couldn't get their car to start, for some reason." She fluttered her hand. "It was an old Chevrolet. After I finally got it started, your aunt Juanita and I drove up to the hospital. So I was telling Dad about my day, things that Mama and Heather and I had been doing—what we had for dinner, and that Mama had let Heather fix cinnamon toast for breakfast. I just felt like I had to keep talking, because I knew that he could understand, that he was in there. And I told him that it was raining that morning, and for some reason, I couldn't get the car to start."

She paused. "And without opening his eyes, he said, 'Sometimes the carburetor gets wet.' That's all he said." Her eyes filled with tears. "Then, later on that night, he died," she said. "That was the last thing he said."

I jumped up to get her a tissue, and one for myself. "Oh, Mom," I said, as we honked our noses. "I didn't mean to upset you."

"I don't mind," she said, dabbing her eyes. Then she gave me a kiss on the cheek.

"I'm glad you asked," she said.

After our talk in the kitchen, my mother introduced yet another family tradition to our already long list. At the close of our holiday dinners, she had me ask one of my questions ("What skill do you secretly wish you had?" was a recent one), and then each family member would give their answer. How else would I have known that Rob harbored a secret dream to play the banjo?

Gay Icons: A Homo Run!

If your artist has a sizable gay following, sign right up. First of all, what's not to like about Cher, or Madonna, or Kylie Minogue, or Elton John? That said, even if you're not an admirer, anyone who carries on the old-fashioned idea of being an "entertainer," the sort of person who nearly drops dead onstage from trying to hold your attention with singing, wisecracks, half-naked backup dancers, and seventeen costume changes, is at least worthy of respect.

At *Rolling Stone*, I quickly became the magazine's rainbow connection. The hipper writers weren't at all interested in spending an afternoon with the Pet Shop Boys or Sandra Bernhard, but I was. Give me a gay icon any day of the week over some shambling hipster who mumbles about his band's integrity. There are exceptions, of course, but for the most part, they are interview gold. Being dull is unthinkable, particularly for veteran performers like Cher, who will reliably turn it on, even if she's giving her fifth "exclusive" of the day.

She won't get irritated if you're the two hundredth person to ask about the wig room that traveled with her from city to city on her endless farewell tour. She knows exactly what you're after, and she delivers. Her theory as to why she is beloved by gays? "Because I'm cool." Once I asked her what the most decadent night of her life was. "I would say from 1974 to 1981," she said. Yes! Nothing was off-limits—or, on the rare occasions that it was, she would tell you to go to hell in a wonderfully quotable way. I freely asked her about the oft-repeated rumor that after she met Gregg Allman, they went to bed for fourteen hours. (No: She doesn't sleep with people on a first date.)

Your garden variety gay icon simply goes that extra mile, whether it's trashing their fellow artists (thank you, Elton John!), being entertainingly moody or unstable, or discussing their downward spirals in vivid, horrific detail. Stevie Nicks—saluted every year in a New York drag extravaganza called "Night of a Thousand Stevies"—likes to tell the account of her visit to a doctor who informed her that she had snorted so much coke in her lifetime that if she did it just once more, the tiny piece of tissue that remained in her nose would whoosh straight up to her brain and kill her on the spot. You can pack up your tape recorder and walk out the door after a gem like that.

Boy George dependably did all of the above when I stopped by his New York hotel room to chat about his autobiography. He flung open the door wearing a pair of red Chinese pajamas and a billowing robe. "Would you mind talking to me in the bathroom?" he asked. "I'm putting on my makeup." Instant setting! Scattered around the counter were dozens of lipsticks, glittery eye shadows in every possible tint, and pencils. I sat obligingly on the toilet next to him and watched as George carefully shaded his eyes with five different colors, dispensing some useful tips along the way (remember: Pat on, rather than rub, your under-eye concealer).

As he applied lipliner, he told me about his first experience with heroin (the next day, he alternately slept and threw up, sweating and crying), talked about shunning the chance to talk to Keith Richards while on a vacation, saying he looked like a ravaged baboon, and claimed that Gavin Rossdale of Bush was the boyfriend of his drag-queen pal Marilyn in the eighties. Then his mood abruptly shifted, he tired of me, and I was ushered out of his hotel room. "Thank you very much for going," he said, laughing, as the door cruelly banged shut.

Even though most of these subjects can be trusted to deliver, it is always necessary to take precautions. Before an interview, I consult any gay friends who might be serious fans and harvest their spectacularly detailed questions. Then, on the slim chance that my subject is having a dark day and is unable to entertain, I can loosen them up. "Before we begin," I will typically say, "I have a few insanely obscure questions from your more obsessive gay fans." Inevitably, they will perk up at questions that they haven't been asked in years, if ever. When Cher's energy flagged (just for a moment, mind you), I announced that it was Obsessive Gay Fan Time and pulled out a query about her 1978 television special (the one that culminated in a spectacular song medley performed by her, Dolly Parton, and the Tubes), followed by a question about a song from her one-album rock band Black Rose called "Julie" (sample lyric: "Julie, oh, Julie, you're a liar, bitch"). She revived. We moved on.

Before a chat with Christina Aguilera, I conferred with a friend who once played Aguilera's self-affirmation tune "Beautiful" on repeat for an entire weekend.

"While she was on the *Mickey Mouse Club* I think she starred on this one series on the show called *Emerald Cove,*" he said. "It was a soap opera, ask her about some of the plotlines, that'll get her going." He paused. "I notice her eyebrows are a little thicker lately, ask her what's up with that. Oh! And if you listen to that track "Singin' My Song," there's this vocal effect that she does near the end, it sounds like a birdcall, what is it? And at the end of the song you hear her talking with someone named Wassim, who is that?"

I furiously wrote.

"She lived in Japan when she was four, I need to know what part of Japan. And why."

As it turned out, Aguilera needed no contingency plan. She was in a perfectly fine mood, and livened up on her own accord when I asked her, randomly, when the last time was that she really got angry. To my giddy delight, she declared a jihad against some of my fellow journalists, first denouncing a duplicitous writer. ("He was a prick, everything was just blatant lies, just personal bashing, and if he felt that way he should have just told me to my face, you put your whole heart and words into the hands of this person, he should really be fired and taken out of the whole job of being a journalist.") Then she ripped into another writer who was pleasant to her face and then wrote an article that was "catty, stupid, and horrible, don't do me like that, it's not cool, I definitely put a thing up on my Web site and I had everybody call her; seriously she had to change her number after a while."

And here's the thing: I sided with Christina. I never understood why there was so much animosity toward her. Writers should have been worshipping her. Were we not drowning in a sea of boring, prefab celebrities? Give me a gal who boldly speaks her mind, who isn't afraid of a fight, who wears a giant Afro wig one day and assless chaps the next. Christina Aguilera is the new Cher, if you think about it. Let us be glad. I never found her to be "difficult," but then, the difficult ones never are. It's the poisonous "America's Sweethearts" that you should avoid.

The day that my profile came out, Aguilera sent me a huge bouquet of flowers, which never, ever happens. "Thank you for giving me one of the most honest interviews I've ever had. It was a pleasure. Love, Christina A."

18.

As my life took a gentler turn, I started to check in, sometimes on a daily basis, with my mom. I marveled: When did we start having so much in common? Had it always been this way and I hadn't noticed? We would talk, sometimes for hours, about books, travel, politics. I had always been the classic independent eldest child, but I found that I liked to phone up and run things by her. She always had a strong opinion. Never from my mother did you hear *I don't know* or *Hm, that's a tough one.* She just jumped in.

In turn, she would phone me when she was driving around and doing her errands. "I'm on my way to Lord and Taylor," she said, calling Friday morning. "Whoops, hang on. I see a cop. I'm not supposed to be on the cell phone, so I'm going to put it down for a minute." Pause. "Okay, he's gone. So I have to get an outfit for a dinner at the Power Boat Club. I'd rather not spend the money, but what can you do. Listen. I was thinking of something for you to pitch."

My mother wrote a column for her local paper, so she was constantly on the hunt for story ideas, some of which she passed on to me. "You know how there are a lot of dwarfs around?" she asked.

I tried to absorb this. Were there a lot of dwarfs around? Was the tristate area lousy with dwarfs? Perhaps. Perhaps.

"Well, I was thinking: Where do they go when they want to have fun? I'll bet you that in New York there is some sort of *dwarf nightclub.*" Already I was preparing my southern imitation for Heather: Ah you *awayah* of the pre*pon*derance of *dwarfs* among our *pop*ulace?

"I mean, I'm assuming that dwarfs want to go to a place where they don't feel self-conscious. Right? New York has everything, so surely it has a nightclub for dwarfs. Why don't you do an exposé on it?"

I considered the existence of a dwarf nightclub that had somehow eluded the attention of the New York media. "Well, there's the problem of height," I said carefully. "An exposé involves going incognito, which I couldn't do unless I put shoes on my knees."

"Well, you don't have to get smart."

"No, I'll look into it," I said.

"So what are you working on?" she asked.

I told her about my latest back-and-forth with Rosie O'Donnell's publicist. *New York* magazine wanted me to write a sizable, in-depth (i.e., lucrative) feature on her, but her publicist, burned by a torrent of bad press after Rosie's court battle with her magazine publishers, was skittish. I felt sleazy trying to reassure her that despite *New York*'s reputation for the occasionally blistering profile, I wasn't looking to take Rosie down. I actually thought Rosie was funny, and I loved the work she did for charity. That said, if I passed muster, the publicist was urging me to go on R Family Vacations, a cruise for gay families organized by Rosie's partner, Kelli.

I informed my editor that I really wasn't a cruise person. "Are you kidding?" he asked. "That's the best scene, ever."

After a few more conversations with the publicist, I was told I made the cut. Still, she was wary.

"I have to proceed gently," I said to my mother. "I guess going on the cruise will help, although I'm dreading it."

"Dreading it? That sounds like a hoot," said my mother.

I was about to protest, but then it occurred to me: My cheery mom from the suburbs would be just the secret weapon I needed. Why didn't I start taking her with me on my interviews? She had retired, so she had more time. For this story, she would be particularly helpful: As my father liked to point out, Gays Love Your Mother. She was fun. She wore big jewelry. She got a little raucous after a few Pinot Grigios.

"Ma," I said. "Pack your bags."

We were to leave New York Harbor, sail for two days and nights, then disembark in Halifax, Nova Scotia, while the ship went on to Provincetown, Martha's Vineyard, Boston, and then back to New York. In the days leading up to the cruise, my mother called with questions. Could I use my press credentials to cut ahead of the registration line? Would she need to pack formal wear, in case there was a dinner with the captain?

"I think the dress code will run along the lines of elasticized shorts," I said. "And you may see some denim-on-denim ensembles."

On the day of the cruise, my anxious father dropped my mother, pert and ready for adventure, off at the port. Wheelie cart: check! Revlon's Pink in the Afternoon lipstick: check! Oversized sunglasses: check! Sleeping pills: check! We took our place in line and watched all of the same-sex parents attempt in vain to control their frantically excited kids. After check-in, we got our commemorative photo taken as we smiled together inside of a life preserver.

Then it was off to inspect our room. "Ooh!" said my mother, grabbing the Daily Events calendar and scanning it. "Well. We have to play bingo, although I'm not lucky. Your aunt Norma always won at bingo. Look! There's a drag-cappella show tonight. What do you suppose that is? Hm. Surrogacy and Egg Donation, I guess that's a workshop. Ooh, eight of the restaurants are free. I know. Let's eat two bites of an entrée and run to the next one! And listen to this! Susan Powter is running the yoga class! Remember, the one with the blond crew cut? We had her book, what was it, *Stop the Insanity*?" She frowned. "Is she a lesbian?"

"I'm not sure," I said.

Mom put on a fresh coat of Pink in the Afternoon and we raced out to the pool. A Pet Shop Boys song was blaring, and everywhere there were kids of different hues splashing and laughing while their doting parents photographed and filmed their every move. My mother sidled up to two white guys who were fastening water wings on their tiny Asian daughter. "Look at her little shoes!" my mother said. "What a beautiful girl." Her parents beamed.

"I'm going with you for all interviews," she announced as we ran to a cooking demonstration by Oprah Winfrey's chef, passing Cyndi Lauper, Wednesday's entertainment, in the hallway. Because Rosie was involved, the onboard talent was of a higher caliber: Tony-winning stars, for instance, were the attraction for Rosie's Variety Hour, the nighttime revue held in the Stardust Lounge.

After joining the stampede for the Chocolate Buffet, we met Rosie's wife, Kelli, and Kelli's business partner, Gregg, for an interview. They told us that with gay parents, there are no accidental pregnancies. These parents often wait, anxiously, for years. They want their children, badly.

My mother dabbed her eyes.

After we left the conference room, Mom hugged me. "I started getting a little emotional in there," she said. Two staffers passed us and exchanged the briefest of amused looks.

"I think we confuse people," said my mother.

"They don't know if you're my mother or my sugar mama," I said.

"Hey!" she said. "Let's interview Susan Powter." Kelli said I could call her room directly, so I did.

"I'm just about to take a nap!" she boomed. "But! I'd love to chat! I have a LOT going on!"

In an hour, she phoned. "Boy, am I refreshed!" she cried. "Let's meet by the pool!"

She strode up to us wearing a tank top that said Big Dyke.

"I guess that answers that question," whispered my mother.

Black and pink dreadlocks spilled down her back. A tattoo snaked around her stomach. She marched up to my mother. "How old are you?" she demanded.

"Sixty," said my mother. That was not strictly true. I let it pass.

"Well, you look amazing!" she said. "I'm a forty-seven-year-old menopausal woman, what do you think of that! Let me tell you about my new book, *The Politics of Stupid*!" She sat down at a table as the families around her tried not to stare. "I hate refined white sugar, refined white flour, and refined white men! Oh, and my latest video is also out, it's called *Trailer Park Yoga*!" She stood up. "I love the vaginal energy in here!" She waved her arms. "Am I right, Mom? *LET*'s get in *TOUCH* with our *VAGINAL* energy!" She leaned over and plopped her ample bosoms onto the table, where they spilled out of her tank top.

Forty-five minutes and many mentions of her new book later, we were on to our next interview, a single gay mother who had adopted seven special-needs kids, all of whom bunched anxiously around us as we talked to her.

"I wish I had a lot of money to give her," said my mother as we ran to the casino. "Jesus God, that is tough work." We passed two large ladies who occupied most of a hot tub and wondered if they were the same big gals who reportedly attended last year's cruise to the Bahamas. They were straight, but they showed up because they got tired of the stares and giggles and figured that on this cruise, they would be accepted.

Finally, we had an appointment with Rosie. She occupied a large suite that looked out over the pool ("I love my lesbian fatties in their bathing suits!" she announced later at her variety show, to thunderous applause).

Rosie was wearing bike shorts and a fleece pullover. She gave us both a hug. Unlike most celebrities, she didn't have that invisible do-not-cross line. She told us that Melissa Etheridge was in the suite next door with her two kids.

"I know her!" I burst out idiotically. "I broke the story in *Rolling Stone* that David Crosby was the surrogate father of the kids! I had to carry around that secret for eight days before the interview."

My mom poked me. "I want to meet Melissa," she announced. The woman was veering out of control.

As Rosie's kids ran in and out of the suite, she talked about how difficult it was to censor herself, which, of course, I loved.

"Someone should censor Tom Cruise," my mother said, referring to his attack on Brooke Shields. She looked over at Rosie to see if she'd take the bait. She certainly had been a vocal fan of his on her talk show.

"Mom!" I said. Then I looked at Rosie, too.

She went off on a satisfying rant against Cruise while I silently vowed that from here on in, I was bringing my mother to every single interview. She could ask the painful questions, while I could pretend to be horrified and admonish her. Then we would wait expectantly for the celebrity's answer, because what kind of monster would bark "No comment" to someone in a pink sweater who has a Master Gardener's certificate and does volunteer work at Project Self-Sufficiency?

"I like Rosie," my mother pronounced afterward. "And I didn't know if I would. But I really think she's real, and unpretentious. Warm. You know?" We didn't have much time to talk, because we were due to interview a pair of women who got a chilly reception from the residents of their small Colorado hometown. We met them in their room.

"We saved for this cruise all year," one woman said haltingly, the tears slipping down her cheeks. "For one week a year, my girlfriend and I don't get stared at. Our daughter sees other kids just like her." She wiped her eyes. "We just feel safe here."

I heard a robust sniff behind me. It was my mother.

"You're very brave," she choked out. Blossoming under my mother's sympathetic gaze, the woman poured out her story, while I busily took notes.

The next day, we disembarked at Halifax. "Well, that certainly opened my eyes," Mom said.

As we walked down the gangplank, I heard a commotion at the entrance of the ship. It was the woman with the special-needs kids. "Come on, everybody," she was shouting as the group made their boisterous journey down the ramp. "Let's stay together."

I squinted up at them. "It's the kids with the special needs," I said.

We looked at each other.

"Run," I said.

A Reminder, After Years of Jangled Nerves, That the Famous Are Not So Different from You and Me

Steady. Steady, now. Bruce Springsteen was standing right next to me. The man who provided the sound track to a few decades of my life was a mere three feet away. We were both in the audience watching his wife, Patti Scialfa, perform onstage in a New York studio for a television special. I was to interview her the next day.

I practiced a line, just in case he caught my eye. *She's great, isn't she?* Emphasis on "she." No, no, emphasis on "great." But as Patti ran smoothly through song after song, he never did look around. Instead, he gazed at his red-haired wife with pride and delight, a little smile playing over his face.

The following day, I met Scialfa in a recording studio on the West Side, where she was putting the finishing touches on a couple of tracks. I had a mild knot of nerves in my stomach, but they vanished immediately as she ran over and warmly said hello, catching both of my hands in hers. "I'm from New Jersey, too!" I cried, before realizing that she must hear that two hundred times a day.

She was funny and friendly and utterly unaffected. When the subject of her age came up, she moved her face under a nearby light. "I'm fifty," she said matter-of-factly, lifting up her bangs so I could see her forehead. "This is what fifty looks like."

There were so many things that I wanted to ask her. She told me about her girlhood at the Jersey Shore, where she hung out at the beach and cruised the streets in a Firebird and cut classes at Asbury Park High School to go into the city. At fourteen, she joined her first band, which I rejoiced to discover was called Ecstasy. Until recently, she said, she had forgotten that at the age of fifteen, she had called Springsteen, who lived a town over, to audition for his band. (He told her, firmly but kindly, that she was too young.)

The time slipped distressingly away as she reminisced about her days as a busker on the streets of New York, and her subsequent career as a singer with the E Street Band. Now she and Springsteen live with their three children in a nineteenth-century farmhouse in Rumson.

She laughed when I told her about the Jersey Boast. If you lived in Rumson or nearby, you were not issued your driver's license unless you could do the Boast, a story of your personal encounter with Bruce and Patti that ended with the proud declaration that they were Just Like Us, i.e., "I saw Bruce and Patti at the grocery store/diner/dry cleaners, and they were so freakin' down to earth. They were buying trash bags/talking to my sister's cousin/eating pancakes while they sat in a booth, just like *everyone else*."

As she continued to talk, I found that my teeth were dry and my cheeks hurt from smiling, and still, I listened, rapt. Before Scialfa's publicist pried me away, I asked her about her home life with Bruce and the children. She said that the kids were grossed out when the two of them kissed in the kitchen, but that she told them, "Hey, you're going to be happy one day when you look back and know your parents really loved each other."

As she relayed that story, I felt the smallest, smallest pang of envy.

19.

Saturday afternoon and my father was on the phone droning about cashing in my pension plan early.

> You're looking at a monthly number when you turn sixty-five and it was developed by having a cash amount in there, you put some in and your employer does, see, and whatever that amount is, by the

I scanned the television section of the paper. Should I stay in and watch that Ric Burns documentary? I could pop a big bowl of popcorn and climb right into my robe.

> time you're sixty-five, they take the number of months they expect you to live past sixty-five, so that's, let's say one hundred eighty months, and they divide it by

No more milk? I feel like I just bought some. I think I have whipped cream somewhere in here. I don't know, is that weird to put in coffee?

> so your decision is whether you want to take the money out and invest it yourself and have better growth than what you're doing, or leave it to their wisdom to invest it and elect to take the monthly payments

On the other hand, I could go to that party at Nathan's, but I'm so tired. By the time Saturday night rolls around I just want to sleep. It's funny. Now that I never go out, I have no energy anymore. What to do? Heather would tell me not to turn down any invitations, but she doesn't like to go out, either.

> the downside is if you should die when you're sixty-seven, you're screwed. If you live to eighty-five and they project you're going to live to eighty, you win. So what I'm saying is, if you invest wisely it will be at least as good as theirs or better.

He seemed to be finished. "I will, Dad!" I chirped.

"You will what?" he said. "Cash out your pension or keep it in there?"

Oops. Wait, he said something about investing wisely. That must mean he wanted me to cash it out.

"I'll cash it out, then," I said.

"Well, I think that's a good idea."

I hung up and got into my robe. There, it was decided. I was staying in. I brought in a bunch of pillows from my bed and piled them onto the couch. Maybe I'd take a bath and read a book. It was raining out, which looked to be a Patricia Highsmith kind of night.

I had been reading for a few hours when I abruptly put my book down. Uneasiness was creeping in. I was beginning to delight in my own company so much that it was getting harder for me to emerge from my apartment. I sat up. Well, maybe I would go to the party. I could swing by for one hour, and then I could leave. I could be back here by ten. Nathan worked at *Rolling Stone,* so at the very least, there would be plenty of people there who I knew.

I roused myself, threw on a dress, and swiped on a little lipstick. I just didn't feel like doing the whole production. Then I quickly went downstairs and hailed a taxi before I could change my mind.

There is something so gratifying about walking into a party alone. I didn't spot anyone I knew, so I made my way to the bar. "I don't feel like alcohol," I told the bartender. "Is there something you can concoct that's good without it?"

"Sure," he said. "I'll do a pineapple juice, maybe some coconut cream, a little ice." He put a cherry on top, and a slice of pineapple, and presented it with a flourish.

"Very festive," I said.

A man appeared at my elbow. "Remember me?" he asked.

I squinted at him. The place was lit with candles, so I couldn't see that well.

"I'm afraid I don't," I said.

"It's Tom," he said. "I had dinner with you and Casey a few months ago."

"Oh," I said. "Right." I dimly remembered him mentioning that he had gone to college with Nathan.

"Can I get you a drink?" he asked. I held up my Carmen Miranda special. "Ah," he said.

A group of *Rolling Stone* staffers rolled toward me and descended, chattering and laughing. Tom drifted away.

"Who was that guy?" asked my friend Susan. "He was cute."

"You think so?" I said. "He's really shy." I sipped my drink. "To the point where it's kind of a strain to talk to him."

"What's wrong with shy?"

I shrugged. "Nothing, I guess."

The group was making noisy plans to go see a band after the party. Chavez was playing at a club downtown. Did I want to come?

"I'll think about it," I said. Hello, couch. Mama will be home soon.

Just as the group moved on, Tom reappeared. "I heard you talking about Chavez," he said. "Do you mean the indie band Chavez, or the farmworkers' advocate Cesar Chavez?" I stared at him. "Maybe you meant the Venezuelan *caudillo* Hugo Chavez? Or was it the Mexican heavyweight boxer Julio Cesar Chavez?" He smiled. "I'm just trying to impress you," he said. He did a little bowing motion with his head, or maybe it was an actual bow.

"Sadly, we were talking about the band," I said. "But I'm not going. I'm actually heading out shortly."

"Well, at least have one more drink with me," he said. He sounded breezy, but as a closet shy person, I knew by the way he kept rocking back and forth on his heels that it was an act. It was touching, somehow.

"Okay," I said.

"So why are you going home?" he asked.

"Oh, nothing exciting," I said. "I want to watch a Civil War documentary. I'm a sucker for that stuff. If there's a lingering close-up of a daguerreotype and the sound of a lone fiddle, I'm there."

He pretended to do a spit take. He was much goofier tonight. Maybe he was bombed. "You're throwing over Chavez for the War Between the States? I thought you were a Rock Chick."

"No, no," I said. "I'm a complete sham. I also taped a documentary about the Great Plague, so it's going to be quite the action-packed evening." I fought a crazy urge to invite him along.

"I've been accused of being a young fogy myself," he said. "Let's just say that my favorite young author is Tom Wolfe." He laughed. "And my last crush on a hot young actress was Veronica Lake."

"I just watched *Sullivan's Travels* the other night," I broke in.

He nodded vigorously. "I saw that, too. Part of 'Preston Sturges Week,' right? Do you know that the last part of her life was really strange?"

"She was bartending in some place in Midtown, and she appeared in some trashy B movie. I can't remember the name."

"*Flesh Feast,*" he said. His smile faded a little and he cast a despairing glance around the room. The music throbbed. "I hate straining your voice to make conversation you barely remember the next day," he said.

I raised an eyebrow at him.

"Not that I won't remember this one," he said quickly.

We chatted for a few minutes as I finished my drink.

"Well, I'm going," I announced.

"Why don't I get you a cab?"

I hesitated.

"It's raining out, you know," he said. "I brought an umbrella."

"Okay," I said. He hustled off to get his coat.

As we stepped outside, a twentyish guy who was lingering by the door darted in front of Tom. He was dressed for a night out: hair gelled into whorls, cell phone in hand, shiny black pants, and, although it was well after midnight, orange-tinted sunglasses.

"Hey, buddy," the guy said urgently. "Do these glasses match the shirt?" He opened his jacket to display a white shirt with purple stripes.

Tom stared at him, perplexed. "Sorry?"

"Do the glasses match the shirt?"

Tom studied the guy carefully. "Sure, I mean, they don't . . . they don't *not* match." He flapped his hand helplessly.

"Thanks, man."

As the rain hit us, Tom quickly opened his umbrella. It occurred to me that I had never gone out with a single person who thought to bring an umbrella anywhere. I flashed on all of the times I was caught in the rain after stumbling out of parties.

"Here we go," he said, holding the umbrella over my head. He put his arm around me. "I'm not making a move, I just don't want you to get wet. Although I would like your phone number."

I sighed. "Why don't I take yours?"

"All right, then," he said, fishing out his wallet and producing a card.

"Thanks," I said, taking it. "Listen, I should get a cab."

He smiled. "Let's walk a little farther. Just indulge me. You should probably know that I came here tonight hoping you might be here."

I looked up at him. He really did have the kindliest expression. Why had I not noticed how blue his eyes were? And it was so cozy underneath the umbrella. I tried to recall why I had rejected him so quickly. Hazily, I remembered that he lacked all of the hipster totems that had usually attracted me. He wasn't my "type." But what, exactly, would that be? Noble failures? Substance abusers with muttonchop sideburns? I had told Casey that he was too quiet. Maybe I had thrown myself into being a New Yorker with a little too much force, joining the herds that trampled over the introverted to flock around the ones who screamed *Look at me!*

"You know what?" I said, smiling up at Tom. "I would love to walk."

"Good," he said. We started down the street, talking so intently that before I knew it, I was practically on my block. At some point as we walked—when, I could never precisely remember—I had slipped my arm through his.

> I can't explain it. He just seemed very familiar to me. I had the pang that Cher said she had when she met Sonny Bono and Rob Camiletti. She says that the time that she met those two guys, the rest of the room went dark.

I stopped on the sidewalk and faced him. For the first time in a long while, I found that I wasn't plotting an escape, or assembling a careful armature of jokes and clever anecdotes. I was completely comfortable. Relaxed. I was—well, I was happy. "I just realized how little I actually know about you," I said. It suddenly seemed important that I should. "I think you mentioned that you lived in Brooklyn, right? Where are you from, originally?" I thought of my joke to Neferlyn the psychic that everyone I met was from Long Island or New Jersey. "Let me guess," I said. "You're from Long Island."

> I see that you will meet a man from the Midwest who will give you his heart.

"I'm from Chicago, actually," he said. "Why?"

ACKNOWLEDGMENTS

Key Club, Spirit Club, Yearbook Committee,
1984 Senior Superlative: Class Clown

Two roads diverged in a wood, and I
I took the road less traveled by
And that has made all the difference

—ROBERT FROST

Thanks to my incredible editor Jill Schwartzman, you are the BEST! Also thank you to Dan Conaway. Even though you transferred schools, you are 2 good 2 be 4gotten! A special thanks to my new buds Jonathan Burnham, Kathy Schneider, Tina Andreadis, Clare McMahon, Carl Lennertz, Tavia Kowalchuck, Stefanie Lindner, Kate Pereira, Sandy Hodgman, Beth Silfin, John Jusino, and everyone else at HarperCollins! You guys rule!

I am sooo grateful to David McCormick, the coolest agent ever, and to my friend Bob Love for hiring me (thanks for making my freshman year the best ever!). Also thanks to my former crew at *Rolling Stone* for all the awesome times: Will Dana, Karen Johnston, Joe Levy, Mary MacDonald, Stu Zakim, and Rob Sheffield.

Special thanks to Jann Wenner. You rock!

A big hug to Julie Klam (Best friends forever!), Lisa Wagner Holley, Susan Kaplow, Tracy Olmsted (party at the Shore!), Rob Stella, and Patrick Williams, as well as Tina Exarhos, Judy McGrath, Karen Infantino, Marlene Rachelle, Sheree Lunn, and Lou Stellato at MTV. (Lou, how much candy have we eaten together over the years? Don't answer! Ha ha!)

A lifelong thank-you to Mom and Dad. Love ya tons! Sorry about all the parties when you guys were away! And to Tom Vanderbilt, I love you "always and forever."

And the most special thank-you to Dinah and Heather, the greatest sisters EVER, my first friends.